Technologies of Inclusion

Gender in the Information Society

Technologies of Inclusion

Gender in the Information Society

Knut Holtan Sørensen
Wendy Faulkner
Els Rommes

tapir akademisk forlag

ISBN 978-82-519-2846-5
ISBN 978-82-519-2848-9 (pdf)

Layout: Type-it AS, Trondheim
Cover Layout: Bjørg D. Wik, Tapir Academic Press
Paper: Munken Lynx 90 g
Printed and binded by: AIT Oslo AS

Publications in *Science and technology series* are peer reviewed. Editors for the series are:
Knut H. Sørensen (coordinator)
Per Østby

Tapir Academic Press publishes textbooks and academic literature for universities and university colleges, as well as for vocational and professional education. We also publish high quality literature of a more general nature. Our main product lines are:

- *Textbooks for higher education*
- *Research and reference literature*
- *Non-fiction*

We only use environmentally certified printing houses.

Tapir Academic Press
NO–7005 Trondheim, Norway
Tel.: + 47 73 59 32 10
Email: post@tapirforlag.no
www.tapirforlag.no

Publishing Editor: laila.andreassen@tapirforlag.no

Foreword

This book is the outcome of a European study entitled 'Strategies of Inclusion: Gender in the Information Society' (SIGIS). This study was the largest single research effort ever undertaken on the subject of gender and information and communication technology (ICT), and a very comprehensive one at that. It represented a path-breaking shift from existing research, with its explicit and unifying focus on the inclusion of women into the Information Society. By contrast, most research on (various) social relations in the Information Society has tended to emphasise digital divides rather than digital inclusion, with a focus on how ICT and new media contribute to the exclusion of women and other vulnerable groups. And most research on gender and technology has tended to address how hegemonic masculine identities are symbolically reinforced around technology. We sought instead to understand better the processes by which gender(s) might be changing around new digital technologies.

To this end, the SIGIS project produced 48 discrete studies, encompassing 30 strategies to include more women in the design and use of ICTs across five European countries. This sizeable empirical effort enabled us to examine a range of different types of inclusion strategies – including training initiatives, design of ICT products and other efforts to increase women's use of ICTs as well as various measures intended to increase the representation of women in ICT. The research also enabled us to capture considerable diversity in terms of gender-technology relations, encompassing a range of different ICTs and a range of different social groups and settings. In this way, it promotes a more diverse, dynamic and heterogeneous understanding of the ways in which gender and ICT are being co-produced across a wide range of activities and settings. Moreover, the research provides much needed positive and practical insights into strategic interventions to 'democratise' the potential benefits of the Information Society by making digital technologies attractive to girls and women.

We hope that, as the product of this research effort, our book will prove useful both to practitioners and policymakers, and to scholars and students interested in new perspectives on gender and ICT and on digital inclusion – including those coming from feminist studies, media studies and science and technology studies, as well as others from informatics and the social sciences. Before moving into the substantive content and contribution of the book, a few words are warranted about how it was produced.

The SIGIS project was a profoundly collaborative effort. It brought together 23 gender and technology scholars from five research centres located in Ireland (Dublin City University), Italy (Studio Metis), the Netherlands (University of Twente), Norway (Norwegian University of Science and Technology, Trondheim) and the UK (University of Edinburgh). In alphabetical order they are: Wendy Faulkner, Leopoldina Fortunati, Helen J. Gansmo, Deirdre Hynes, Aphra Kerr, Tine Kleif, Vivian A. Lagesen, Lisa Lee, Roberto A. de Luca, Carol MacKeogh, Anna Maria Mangelli, Hege Nordli, Barbara O'Connor, Ellen van Oost, Nelly Oudshoorn, Paschal Preston, Els Rommes, Marcelle Stienstra, Kristin H. Spilker, Irma van Slooten, James Stewart, Knut H. Sørensen, and Robin Williams. Regular meetings were held to develop the design and analysis of the 48 studies and, towards the end of the project, to reflect on cross-cutting themes as these emerged. Robin Williams did an excellent job as coordinator and manager of this complex project, and we are very grateful for his efforts. The study was funded through a grant from the European Union's 5th Framework Programme (IST-2000-26329).

This book represents a continuation of the SIGIS project, and so rests on the collective efforts of the whole SIGIS team. We have drawn heavily on the focused study reports of all our colleagues, without which this book simply would not exist. These are listed, along with a short summary, at the end of the book and cited in square brackets throughout the text. The writing job we took on was therefore both synoptic and synthetic of this wider pool of research and reflection. At the same time, our own further reflections on the combined material from the project have progressed substantially on the cross-cutting analysis started by the wider SIGIS team. Whilst some new ICTs and inclusion practices have emerged in the interim, we believe the broad conclusions of our effort – both theoretical and practical – remain salient.

We would stress that this is a book rather than an edited collection. We believe it represents a coherent account and analysis of the topic, which step-by-step develops and elaborates our main findings, arguments and conclusions. The book has been produced through a profoundly collaborative and iterative process, with each of us working on each of the chapters several times as we reflected inward (on the analysis of each inclusion strategy studied) and outward (on how this might inform our deeper thinking about digital inclusion and about gender and ICTs). Three of the chapters originated from other SIGIS team members – Helen J. Gansmo, Vivian A. Lagesen, Lisa Lee – and a fourth built on inputs from James Stewart. These contributions are acknowledged in the authorship of the chapters. Also acknowledged is the fact that for most of the chapters one of us acted as lead author, notwithstanding the substantial contributions of the other book authors. As a native speaker, Faulkner carried the largest role in terms of textual and linguistic editing.

The product of this effort is a book that makes a unique, and we believe crucial, contribution to our understanding of how to further gender inclusion in the Information Society, and at the same time intervenes critically in debates about gender-technology relations. We trust you will find it helpful and worthwhile.

Wendy Faulkner, Els Rommes and Knut H. Sørensen
Edinburgh, Nijmegen and Trondheim
August 2011

List of Contents

Foreword 5

1

On the road to the Information Society:
Does gender still matter? 13
A disappearing digital gender divide? 13
The persistence of gender gaps 17
Moving targets? Gender and the Information Society 24
Equal opportunities: Category politics and gender mainstreaming 29
The SIGIS effort 31
Overview of the book 33

2

Changing perspectives on gender and technology:
From exclusion to inclusion 41
Gender and technology: What kind of problem? 41
Exclusion narrative I: A world without women? 44

Exclusion narrative II: A chilly ICT culture 47
An inclusion narrative: The woman communicator 50
A dynamic view of gender and ICT: The co-construction concept 51
Thinking from exclusion 54
Thinking from inclusion 58

3

Reaching for the digitally excluded 63
Local experts and social networks 67
Roles of local experts in digital inclusion 69
The Ardmore Network 73
How digital in/exclusion is gendered 77
One size does not fit all! 83

4

Websites for women: New routes to digital inclusion 87
Commercial women's magazines 89
Femme (Norway) 89
Libelle (Netherlands) 91
iVenus (Ireland) 93
Donna Moderna (Italy) 95
Women's web magazines as 'assisted self-inclusion' 96
Inclusion through on-line interactivity 100
Symbolic aspects of digital inclusion 103
Conclusions 105

5

Fun and play in digital inclusion 109
Previous research on gender, ICT and fun 110
One place to have fun and another to do 'proper' computing... 114
Fun and social computing as a way to break dichotomies? 116
Fun as a gateway to education? 120
Women just want to have fun? 123
Toy and tool, fun and useful, boys and girls 125

6

Inclusion by design? 129
Inclusion of women: Commercial motives? 131
Why include women? 134
Design practices and inclusion methods 136
Designing from stereotypes 137
The I methodology 139
Reflexive I methodology 140
User testing and participatory design 142
Gender inclusive products through gender inclusive design? 143

7

Getting women into computer science 147
Analysing exclusion and inclusion: Women in ICT 147
The WCI: Construction of an inclusion strategy 152
Visibility and diversity: Actions within the WCI 154
Actions targeted at school girls 155
Actions aimed at getting women to apply: The quotas 155
The advertising campaigns 157
Other measures aimed at getting women to apply 159
Actions aimed at getting women to accept their offered study place 159
Actions aimed at retaining women students 160
Lessons to be learnt: Why was the WCI a success? 163
Making positive circles 166

8

Empowering women in ICT through professional networks 171
The networks 174
Being a minority in ICT workplaces 175
Networks of Empowerment 177
Community 178
Professional Development 179
Visibility 181
Networks as empowerment 183
Bottom-up mechanism 183
Transparent, flexible and open 185
Conclusions 188

9

A room of their own? Inclusion through women-centred ICT spaces 191
Women-centred spaces and ICT? 192
Women-only ICT training: A stepping stone to inclusion 196
Networks for women in ICT: A safe haven from men? 201
Women's spaces on the Internet: A virtual room of one's own?. 204
Designing for girls or women 207
Does 'feminisation' work? Critical reflections on women-centred spaces 210

10

The anatomy of inclusion 215
Initiating inclusion 217
Strategic thinking: The challenge of gender stereotypes 220
Inclusion models and inclusion instruments 226
Quantity and quality 230
Technologies of inclusion 235

11

Changing Gender? 241
Co-productions of gender and ICT: The role of stereotypes 243
ICT – a technology of inclusion in itself? 246
Can we go beyond gender? 249

Brief overview of the SIGIS focused studies 253

References to the SIGIS focused studies 273

References 277

Authors 299

1

On the road to the Information Society: Does gender still matter?

Knut H. Sørensen

A disappearing digital gender divide?

"Why can't women be ... like men?" complains Professor Higgins in the musical *My fair lady*. This blunt, politically incorrect statement captures a widespread sentiment concerning the digital gender gap – that everything would be fine if only girls and women could engage with the new information and communication technologies (ICTs) like boys and men do. This is perhaps not surprising. Computers have appeared to be much more attractive to young men than to young women, and it has been claimed that many young women stay away from computers because they consider them to be a boyish or masculine thing. Perhaps for these reasons, there has been widespread pessimism about the possibility of closing the gender gap – not least in relation to efforts to increase the representation of women in computer specialisms, which have proved so resilient to policy intervention for so long. This was expressed dramatically at a seminar in 2004 by a prominent woman professor of computer science who declared it to be "easier to put a man on the moon than to get more women to enter computer professions" (Lagesen 2005:2).

In sharp contrast to this pessimism, it has become commonplace for others to consider such 'women's issues' a thing of the past, a concern that is no longer relevant in the 21st century. While there may be some differences between men and women, such dissimilarities are not really relevant to participation in the Information Society, so the argument goes. The important factors in digital inclusion/exclusion are individual differences in interests, tastes, competencies and skills – not gender. Often, when gender-related differences in access and use of ICTs are acknowledged, it is assumed that these will 'spontaneously' diminish over time.[1]

The presumed 'spontaneity' in this view is market driven. And indeed, some previous research suggests that market actors might generate more promising inclusion initiatives than public policy-makers (Cassel and Jenkins 1998, Spilker and Sørensen 2002). Spilker and Sørensen provide the striking example of a large Norwegian producer of 'feminine hygiene' products that decided to spend some of its marketing budget on a booklet intended to introduce girls to the Internet – an intervention which contrasts with observations of how Norwegian public schools fail to interest girls in computers (Gansmo 2004). Perhaps the best example to support the contention that the gender digital divide is being taken care of by market forces is mobile telephone ownership. It seems that women have caught up with men with respect to this technology, at least in countries with a generally high rate of access. In the US, more women than men own a mobile phone (Castells et al. 2007, pp. 41). A Eurostat survey from 2008 confirms that there are no significant gender differences in access to mobile phone except for the age group 55–74 in some countries (see Table 1).

1 See for example eEurope Advisory Group. 2005. *e-Inclusion: New challenges and policy recommendations* (Rapporteur: Daniel Kaplan), p. 13. (Downloaded 2007-04-4 from http://ec.europa.eu/information_society/eeurope/2005/doc/all_about/kaplan_report_einclusion_final_version.pdf).

Table 1: Percentage of population using mobile phones, according to sex and age, by 2010.[2]

	Age 16–24		Age 25–54		Age 55–74	
Country	Men	Women	Men	Women	Men	Women
Germany	96	98	92	90	79	69
Ireland	99	98	97	98	81	85
Italy	95	97	95	95	83	74
Netherlands	99	99	95	95	85	82
Norway	99	100	100	100	92	91
UK	99	100	97	95	83	85
EU (27 countries)	96	97	93	92	75	69

The case of mobile phone use appears to be in line with the main assumption of diffusion research that over time the uptake of new technologies follows a pattern called the S-curve (Rogers 2003). According to this model, the spread of the innovation is initially slow, because pioneer users are few. Pioneer users typically have a particular interest in new technologies, plus the time, resources and skills needed to find out about innovations and how they may be applied in gainful ways. Then follows a period of more rapid uptake, in which more and more people see the benefits of the new technology. This rapid uptake phase is followed by a stage where appropriation again goes more slowly, because the market niche is becoming saturated. Those who become users at this stage, or who elect to remain 'non-users' altogether, are often people who need to be careful with spending scarce resources such as money and time. Thus, while it is common to observe social inequalities in ownership and use of new technologies in the early life of an innovation, diffusion theory suggests that these inequalities are reduced but not necessarily done away with entirely as the new product becomes cheaper and better known.

From the perspective of diffusion research, then, we should actually expect a diminishing of the gender gap in access to and use of the most popular ICTs until it more or less goes away – as has happened in the case of the mobile phone. Because of the S-curve phenomenon, it is argued, inclusion initiatives are not needed. Images of the pervasive use

2 Source: Eurostat (downloaded 2011-06-17 from http://appsso.eurostat.ec.europa.eu/nui/show.do?dataset=isoc_sk_cskl_i&lang=en)

of new social networking technology such as Facebook, promoted by the news media, appear to confirm this. However, as we shall see in the next session, there are good reasons to be critical of claims about a disappearing gender digital divide when we analyse European data.

Other statistical indicators also caution against unqualified optimism on the topic (see, e.g., OECD 2007). It seems likely that the phenomenon of 'spontaneous', S-curve-like reductions in gender gaps with respect to ICTs is overstated. But does this mean we must embrace a pessimistic stance on the subject? Is the European Parliament Committee on Women's Rights and Equal Opportunities correct in concluding that 'Lack of access to the new technologies and digital illiteracy are increasingly becoming new forms of social exclusion which pose a direct threat to women'?[3] Clearly, as the Committee concludes, there is at least a need for better analyses of the situation.

This book is intended as a response to such calls. It offers a detailed investigation and analysis of diverse strategies to get more women included as designers and users of these new technologies and media. It represents a radical departure from the dominant traditions in the study of gender and technology – above all, through its explicit and unifying focus on the *inclusion* of women in the Information Society, rather than on exclusion processes. In this way, the book combines a policy interest in strategic efforts to reduce the gender digital divide, with an academic interest in how gender-ICT relations may be changing across a wide range of activities and settings. We study what such changes imply for the doing of gender as well as of ICT. As the reader will see, the result of this endeavour represents a challenge to both the pessimism and the optimism surrounding gender inclusion in the Information Society.

The remainder of this introductory chapter sets the policy context for the book. It presents evidence on the persistence of gender gaps with respect to ICTs, to highlight why interventions are still needed if we are to ensure gender inclusion in the Information Society. It then starts to unpack the very concepts 'gender' and 'the Information Society', arguing that both are, in effect, 'under construction'. It also contrasts two approaches to gender equality practices – 'category politics' and 'gender mainstreaming' – since these frame many efforts to include more women in the Information Society.

---

3 *Report on Women in the New Information Society* (2003/2047(INI)), p. 12. (Downloaded on 2007-04-04 from http://www.europarl.europa.eu/sides/getDoc.do?pubRef=-//EP//NONSGML+REPORT+A5-2003-0279+0+DOC+PDF+V0//EN&language=EN).

We then introduce the empirical basis of the book – a trans-European research project that studied numerous diverse examples of digital inclusion strategies – before outlining the remaining chapters of the collection. This project ran between 2002 and 2005. The sceptical reader may believe that studies from this period are outdated, given continuous developments in the ICT area, in particular the emergence of social networking technologies. However, there are good reasons to believe that the inclusion challenges with respect to these new media are quite similar to those regarding access to and use of computers and the Internet. Moreover, the theorising of the dynamics of the gender-ICT relationships that we offer in this book remains valid.

Of course, one should be sensitive to the potentially changing nature of what constitute information and communication technologies. The label ICT was originally introduced into public discourse to replace 'computers' as the object of concern. It highlighted the point that computers were no longer simply computational machines but were becoming tools of both information gathering and communication. In our use of the concept, we include mobile phones as well as new social media, e-book readers and smart TVs. Nevertheless, we would stress that the computer remains the technological centre-piece of the Information Society. The computer is for most people the tool through which they access the Internet, chat, use Skype, download music and movies, do Facebook, etc., even if the mobile phone is also increasingly used for such purposes. This means that computer skills as well as Internet competence continue to hold a critical place with respect to participation in the Information Society. Accordingly, while we do address mobile phones in this book, we are mostly concerned with the ICTs of computers and the Internet.

The persistence of gender gaps

Many nations have for a long time initiated activities designed to achieve a transition to what has commonly been called the Information Society. The concept of the Information Society was originally developed from observations of increasing importance of knowledge and services, relative to agriculture and industry (Bell 1973, Crawford 1983). However, the term has come to cover an increasing range of sociotechnical practices (see below), practices which are diagnosed and understood in a variety of ways (Castells 2001, 2004a, Kumar 1995, Preston 2005, Webster 2004, Wyatt et al. 2000). Nevertheless, the ambiguities and ambivalences present in scholarly discussions have not precluded an increasing popularity of the concept, not least in European Union policymaking.

European Union policies have for some time emphasised the vision of 'An Information Society for All'.[4] Even in the Europe 2020 publication, inclusion issues are still high on the agenda.[5] This is premised on the claim that the widespread use of new ICTs offers a wide range of potential benefits, with huge significance for both society and the individual. These benefits include improvements in economic growth, sustainability, individual welfare and self-expression, citizenship and democracy, social inclusion and empowerment, as well as diverse other benefits flowing from greater access to knowledge and information.[6] Of particular interest to our endeavour, digital inclusion is frequently held out as providing a route to enhanced social, political and economic inclusion for socially marginal groups. In this context, the digitally excluded are seen as doubly disadvantaged: they miss out on the potential benefits of using ICTs, and they are further marginalised because of this. At the same time, the production and use of knowledge (using advanced ICTs) is becoming increasingly important as the basis of what is commonly called the 'new knowledge economy', in a situation characterised by increasing global competition among nations and regions. To reap the potential benefits of the Information Society, it is therefore seen as necessary to combine efforts to promote innovation in ICT as well as in the use of ICT.

In short, digital inclusion is seen as vital to future progress and well-being. Whilst we do not necessarily buy into all of the policy rhetoric about the Information Society, we do accept that the greater use of ICTs can enhance people's lives and their opportunities in all sorts of valuable ways. It is also clear to us that many ICT skills will soon gain the status of necessary prerequisites to citizenship, consumption and employment. These opportunities and resources should, therefore, be available for all.

Many studies have analysed social differences with respect to ICT, establishing the potential existence of a digital divide (for an overview, see, e.g., Selwyn 2004, Norris 2009). Two important observations should be

---

4 See, for example, Commission of the European Communities. 2004. *Challenges for the European Information Society beyond 2005.* (Downloaded on 2007-04-07 from http://ec.europa.eu/information_society/eeurope/i2010/docs/communications/new_chall_en_adopted.pdf). Cf. also World Summit on the Information Society. 2005. *Tunis Agenda for the Information Society.* (Downloaded on 2007-04-07 from http://www.itu.int/wsis/docs2/tunis/off/6rev1.pdf).

5 *Europe 2020 – A European Strategy for Smart, Sustainable and Inclusive Growth.* http://ec.europa.eu/eu2020/pdf/COMPLET%20EN%20BARROSO%20%20%20007%20-%20Europe%202020%20-%20EN%20version.pdf (downloaded 2011-02-25).

6 See note 4.

made from this literature. First, one needs to avoid subscribing to the technological determinist view that increased access to ICTs in itself is useful and may contribute to reducing social inequalities (Gunkel 2003, Sassi 2005). Secondly, empirical analyses need to move beyond the traditional focus on access to engage with issues related to effective use and actual benefits from ICTs (Gurstein 2003, Selwyn 2004, Tsatsou 2011). In addition, rather than seeing the digital divide as a binary of included and excluded, we should consider digital inclusion as a continuum (Livingstone and Hesper 2007, Tsatsou 2011).

How, then, may we measure inclusion into the Information Society? What are appropriate indicators? Given the scholarly disagreements, these are difficult questions (see OECD 2005 for a comprehensive discussion). Pragmatically, however, available statistical information covers four broad categories mainly: (1) access to and use of the Internet, (2) computer skills, (3) higher education graduates in computing, and (4) the ICT workforce. Whilst these categories miss out much that we might value, they do at least provide some basis to judge whether or not we still need to worry about a gender digital divide.

Table 2. Percentage of population in the age group 16–74 using the Internet daily, according to sex, 2004–2010.[7]

Country	2004		2006		2008		2010	
	Men	Women	Men	Women	Men	Women	Men	Women
Germany	37	23	46	33	58	44	67	53
Ireland	14	11	28	22	38	39	48	46
Italy	30	7	32	24	41	30	52	40
Netherlands	n.a.	n.a.	68	54	72	61	81	72
Norway	50	28	66	51	77	67	85	77
UK	36	22	45	33	58	49	68	63
EU (27countries)	27	18	36	27	47	39	57	49

Undoubtedly, during the last decade or so, women's access to and use of the Internet has grown quickly, as predicted by the diffusion perspective. Katz and Rice (2002: 41 ff.) claim that in the United States, the percentage of women using the Internet even has exceeded that of men. Although

7 Source: Eurostat (downloaded 2011-08-12 from http://appsso.eurostat.ec.europa.eu/nui/show.do?dataset=isoc_ci_ifp_fu&lang=en).

the accuracy of that assertion may be contested (Dholakia 2006), differences are small. However, the situation in Europe indicates more resilient gender differences, as is evident from Table 2. The main impression is of course that there has been a rapid increase in the frequency of Internet use. In some countries, we observe reduced gender differences over the period, but the reductions occur primarily between 2004 and 2006. After that time, gender differences remain fairly constant at about 8–10 percentage points, varying somewhat between the countries.

It is mostly surveys about Internet use that form the basis of claims that the gender gap is closing.[8] As Table 2 indicated, however, the results of such surveys are ambiguous. Differences become more marked when one looks in more detail at measures of Internet use. Significantly, we find that men spend considerably more time using the Internet than women, and they do much more downloading of software (Dholakia 2006, OECD 2007). According to Helsper (2010), men have a more comprehensive use of the Internet than women. In addition, measures of Internet skills, suggest an age gap as well as a gender gap: see Table 3.

Table 3. Percentage of population with a high level of internet skills, according to sex and age, 2010.[9]

	Age 16–24		Age 25–54		Age 55–74	
Country	Men	Women	Men	Women	Men	Women
Germany	26	15	14	5	n.a.	n.a.
Ireland	14	8	7	4	0	0
Italy	33	23	18	8	4	1
Netherlands	14	9	10	4	2	1
Norway	31	29	16	5	5	1
UK	25	16	14	5	n.a.	n.a.
EU (27countries)	29	20	14	6	2	1

Strikingly, Table 3 indicates that the level of Internet skills is fairly low, overall. The high level of skills is measured by having engaged in 5 or 6 of the following activities:

8 See, for example, *Benchmarking Social Inclusion in the Information Society in Europe and the US*. Report from the SIBIS project, 2003. (Downloaded on 2007-04-04 from http://www.empirica.biz/sibis/files/WP5_No6_Social_Inclusion_2.pdf).

9 Source: Eurostat (downloaded 2011-06-17 from http://appsso.eurostat.ec.europa.eu/nui/show.do?dataset=isoc_sk_cskl_i&lang=en).

- Individuals who have used a search engine to find information
- Individuals who have sent an email with attached files
- Individuals who have posted messages to chat rooms, newsgroups or an on-line discussion forum
- Individuals who have used the Internet to make phone calls
- Individuals who have used peer-to-peer file sharing for exchanging movies, music, etc.
- Individuals who have created a Web page.

Statistics about computer skills reveal more marked gender gaps. Table 4 shows the percentage of the population across the European Union and in selected European countries that report themselves as having a high level of computer skills, grouped according to age and gender. 'High level of computer skills' here includes having done 5 or 6 of the following activities:

- used a mouse to launch programs such as an Internet browser or word processor
- copied or moved a file or folder
- used copy or cut and paste tools to duplicate or move information on screen
- used basic arithmetic formulae to add, subtract, multiply or divide figures in a spreadsheet
- compressed files
- wrote a computer program using a specialised programming language.

The findings are striking. As we can see, there are major gender gaps in computer skills across all countries and all age groups.

Table 4. Percentage of population with a high level of computer skills, according to sex and age, 2009.[10]

	Age 16–24		Age 25–54		Age 55–74	
Country	Men	Women	Men	Women	Men	Women
Germany	52	26	46	21	19	4
Ireland	41	25	28	23	7	5
Italy	52	37	35	19	11	2
Netherlands	61	37	64	32	34	9
Norway	69	31	58	30	27	10
UK	47	37	42	24	22	n.a.
EU (27 countries)	50	36	36	21	14	4

We also observe a major age gap in this measure of digital inclusion/exclusion, with a much lower share of high level of skills in the 55–74-year-old age group than in the 16–24-year-old age group. The same survey also confirms the unsurprising conclusion that along with age, education level and occupation are the most significant factors-shaping people's ICT skill levels, these factors cutting across gender. From other sources, there is also evidence of differences by ethnic group (d'Haenens and van Summeren 2002; Volman et al. 2005). Nevertheless, the statistical differences between women and men presented here seem robust, even if we assume that men overestimate and women underestimate skill levels. It is possible that the gender gap with respect to computer skills is diminishing as it did with mobile phones. If we compare the 2009 survey with a similar survey from 2006, there is a small reduction in the differences between men and women. At best, however, it will take a long time to bridge this gender gap, especially since it remains so substantial in the under 25-years old-age group. Policymakers and others should be careful not to stake a lot on the spontaneous dynamics of S-curves.

This caution becomes even more pertinent when we look at statistics related to higher education. Here, we find a far greater gender gap, which has not been bridged at all in most countries. It is a striking and lamentable fact that women remain relatively absent from fields such as computer science, software development, and the design of ICT products in most industrialised countries. Some evidence of this is given in Table

10 Source: Eurostat (downloaded 2011-06-17 from http://appsso.eurostat.ec.europa.eu/nui/show.do?dataset=isoc_sk_cskl_i&lang=en).

5. We see that the estimated proportion of women amongst computing graduates ranges from 9 to 34 per cent, with most countries around a quarter. We have added Australia and Korea to get a broader picture. By contrast, there are countries such as Malaysia where the number of women computer science students is equal to that of men (Lagesen 2008, Mellström 2009).

Table 5. Percentage of women graduates in law, life sciences and computing, tertiary level, 2008.[11]

Country	Computing	Law	Life Sciences
Australia	21	57	56
Germany	16	53	67
Ireland (2005)	29	61	69
Italy	16	59	70
Korea	21	38	48
Netherlands	9	62	59
Norway	21	57	71
UK	20	60	52
US	20	49	59

To put this in perspective, Table 5 compares the proportion of women with a tertiary level degree in computing with that in law and the life sciences, both of which used to be dominated by men students only a generation ago. As we can see, a radical change has occurred in these disciplines, yet in computing disciplines women remain a minority, as they do in engineering and the physical sciences (European Commission 2006, OECD 2007). Of course, this minority status in computing education carries over into the labour market. Within the category of ICT specialists, the 2007 OECD report shows that the proportion of women amongst computing professionals in 2004 was less than 25 per cent in most countries, except Ireland where it was only slightly higher (p. 12). The report concludes that:

> [T]he gender gap with regard to ICT continues. There are significant differences between women and men in ICT-related employment, with women having low shares of employment in ICT specialist occupations (e.g. software engineers, IT specialists), and among intensive users of ICTs they are most

11 Source: http://stats.oecd.org/index.aspx?r=354113 (downloaded 2011-06-17).

> heavily represented in office and secretarial occupations rather than professional ones. These gaps tend to persist over time and in some cases increase, unlike the differences between women and men in other areas of employment except management posts (OECD 2007:41).

Koput and Gutek (2010), similarly, find a gender gap in the ICT industry. While women computer graduates enter the industry to the same extent as men, the field does not provide them with the same opportunities as men. For example, the women in the Koput and Gutek study received fewer job interviews and offers.

In sum, available statistical information does not support claims of a disappearing digital gender gap in any of the four main areas discussed here. Whilst there have been positive developments with respect to the proportion of women and men accessing and using the Internet, there remain large differences in the amount of time spent using it. A surprising gender gap also exists in both Internet and computer skills, even amongst young people. Moreover, the gender gap in higher computing education, and among ICT specialists, is very substantial and apparently stable in most (Western) industrialised countries.

The common interpretation of such evidence is to conclude that women are victims of a diverse set of exclusion processes or deficits. Thus, it is highly relevant to ask, as this book does, how these exclusion processes might be remedied? Put more positively, our overarching concern here is to consider what might be needed in order to improve gender inclusion in the Information Society. As a first step to answering these major questions, we turn now to unpack the concepts of 'gender' and 'the Information Society'.

Moving targets? Gender and the Information Society

As we noted above, the concept of the Information Society has been interpreted in various ways. It is not something that exists; rather, the concept connotes a diverse set of visions or 'digital sublimes' (Mosco 2004), which means that it is a kind of moving target. Thus, inclusion is not an invitation to join a pre-established set of sociotechnical practices; it is a 'work in progress', which is both multi-faceted and subject to multiple interpretations. The concept 'sociotechnical' is a very useful contribution from social studies of technology (e.g., Bijker and Law 1992). It tells us that technological artefacts are thoroughly embedded in the social arrangements surrounding their creation and use. The two are not in any meaningful sense separable, since technology is never 'just' technical and society is never 'just' social. We need to understand one in order to understand the other; they are 'co-produced'.

This has some apparent implications for processes of digital inclusion. Most obviously, it means that including people into the Information Society is not simply an issue of diffusion, of acquiring or getting access to ICT artefacts. Appropriating and 'domesticating' a new technology into our everyday lives (see Chapter 2) typically involves changing meanings and practices around the use of the new technology, as well as learning the necessary skills. A set of related and very sociotechnical 'inclusion needs' have to be in place if people are going to go to this effort (see Chapter 3) – not least, people have to be motivated; they have to see some clear relevance or benefit from the new technology. This in turn means that digital inclusion efforts have to involve an enticement to participate in the Information Society if they are to be successful. So how is the Information Society perceived in inclusion efforts? How is it made appealing?

The importance of offering attractive ways of using a technology seems obvious, but tends to be overlooked or taken for granted in debates about digital inclusion. Often the appeal or relevance of computers is taken to be self-evident by ICT professionals and policy makers, and many inclusion efforts are based on the idea that the technology in itself is so important and interesting that no further arguments are needed to entice people to participate. Where inclusion arguments are offered in official discourses about the Information Society, these tend to be dominated by authoritative statements about the instrumental importance of ICT for working and living – that computer skills are a key competence in the labour market as well as a gateway to the Internet.[12] It may be that such messages resonate less well with those who do not participate in the labour market, or who are not already computer-active – including, possibly, quite large audiences of girls and women. This is not to deny the practical utility of many ICTs, nor that access to ICTs and related skills can be very real assets in working life and elsewhere. However, it is important to investigate whether instrumental arguments actually hit home, an issue we return to in Chapters 4 and 5. Is there a need to frame benefits from working and living with ICTs in a different fashion than the instrumental one?

Since ICTs have become ubiquitous in social and economic life, our encounters with these technologies occur in complex interactions between education, work, home life leisure, and increasingly political participation. Accordingly, digital inclusion efforts are targeted on

12 See note 4.

a range of arenas: education, labour market participation, citizenship and consumption. But in principle at least, there is a potentially limitless range of sociotechnical practices and meanings through which people may appropriate ICTs. Since the Information Society is a work in progress, it has the potential to be shaped not only by those who perform inclusion but also by those who become included – another good reason for promoting digital inclusion. For this reason, the book does not understand digital inclusion as a goal in itself; rather, we see the key feature of digital inclusion as being the emergence of a whole gamut of new sociotechnical, ICT-related practices and meanings which are considered attractive by the diverse groups of citizens that make up society.

Emphasising what it is that people choose to do with ICTs, rather than ICT artefacts or even skills, helps to overcome some problematic aspects of debates about the digital divide. First, it enables us to recognise that for some people there may be perfectly good reasons not to use some ICTs (Brants and Frissen 2005, Wyatt et al. 2000) – something that is typically overlooked in the hyper-enthusiasm surrounding calls for digital inclusion. Secondly, it also avoids privileging certain ways of applying such technologies – as, for example, in strategies that idolise the approach to computers often attributed to enthusiastic men hackers (Gansmo et al. 2003a, b). Slogans such as 'An Information Society For All' should be understood as allowing for diversity – considerably more diversity than suggested in the usual statistical indicators of digital inclusion. It is not required that women should be like men or, for that matter, that everyone should be a hacker. Clearly, we need to study to what extent such normative expectations are 'built into' inclusion efforts.

Whilst our empirical focus in this book is principally on the inclusion of women in the Information Society, our analytical framework is very deliberately couched in terms of gender. The concept of gender is, like that of the Information Society, also one that means different things to different people.

Most commonly gender is used (and understood) as a binary concept, referring to presumed differences and inequalities between 'women' and 'men'. We have already encountered this use in the statistical information presented above. This model of gender assumes, first, that there are significant *differences between* the categories 'women' and 'men' and, second, that there is *sameness within* these binary categories. Both assumptions are highly questionable empirically. Aggregate differences between 'women' and 'men', where they exist, typically mask huge differences amongst women and amongst men. This is readily seen in relation to indicators of digital inclusion. For example, there may be very little difference in the computer skills of women and men university students, but huge differences between men students and middle-aged men manual workers. What we learn from such insights is that there are in practice diverse genders in terms of people's identities and practices. The category 'men' encompasses a great variety of men, and a great variety of masculine- and feminine-connoted identities; and the category 'women' encompasses a great variety of women, and a great variety of feminine- and masculine-connoted identities.

For this reason, it is conventional wisdom in social science to conceive gender as necessarily plural – hence the insistence on diverse genders. Gender is also *relational*: in order to understand the lived reality of particular groups of women or men, one has to look at the specifics of their particular social locations (home, family, work, country, life stage, etc.) and cultural activities and values (religion, leisure, etc.). This is evident in the 'intersections' of age, occupation, education and so forth with gender in statistics about ICT use and skills. Time and again in our analysis of various digital inclusion efforts, we were reminded how much 'specifics' matter.

Our objection to thinking of gender as a binary is political as well as empirical. The binary model of gender leads inextricably to a whole series of stereotypes about how/what 'women' and 'men' are – stereotypes which appear to essentialise, as if natural and unchangeable, singular readings of 'femininity' and 'masculinity'. Whilst many gender stereotypes may appear innocuous, even amusing, they frequently serve to reinforce, if not justify, inequalities between women and men (Lagesen 2003). (The common jokes about how only men can read maps are not only inaccurate; they can also undermine women's confidence about being independently mobile.) Often, lurking behind the presumption that 'women' and 'men' are different is an implicit value judgement that one is better than the other. Indeed, dualistic binaries tend to invite a hierarchy, so contributing to unwanted inequalities (Faulkner 2000a).

For all these reasons, we choose to put 'women' and 'men' in quotation marks when we use these labels in the category sense of all women and all men. The scare marks serve to highlight our profound discomfort with over-generalising about either presumed differences between women and men, or presumed sameness of women and of men.

The critique of the binary conception of gender, well founded though it is, nevertheless creates a problem for feminist scholarship and practice. How to provide and manage analytically an understanding of gender that allows one to see sameness and difference at the same time? How to retain gender as a scientific concept without (re)producing essentialist readings and stereotypes? How to acknowledge meaningfully the profound diversity of the categories 'women' and 'men', without lapsing into a liberalist perception that gender does not matter (cf 'We're all individuals')? On a policy level, these concerns translate into the challenge of promoting equal opportunities without assuming sameness, of supporting gender diversity without reproducing unwanted inequalities.

These are questions and challenges feminists and others have grappled with for a long time (see, e.g., Lloyd 1984, Braidotti 1994, 2002, Butler 1990, Fenstermaker and West 2002). And we face them also. On the one hand, we need to keep 'men' and 'women' as analytical categories – in order to study gender-motivated efforts of digital inclusion, and to be able to see and understand how inequalities related to gender are acted upon (and acted out). After all, it is the binary calculation of gender and ICT that reveals the digital gender gaps, which provide the starting point for this book. On the other hand, we have a deep scientific and political commitment to acknowledging diversity and challenging binary conceptualisations of gender.

Our strategy for handling this tension is the one that has been favoured in much feminist scholarship over the last twenty years (e.g., West and Zimmerman 1987). It is to understand gender not as something that is – fixed, universal categories of women/men, femininity/masculinity – but as something that is done. Within this 'performative' concept of gender, the categories of women/men, femininities/masculinities are understood as outcomes, rather than the starting point, of analysis. Within this framework, the focus of inquiry is on the process of gendering, the way particular gender categories are attributed and mobilised. This focus obliges us to address many dimensions and dynamics of gender – including symbolic images and cultures as well as practices and identities. It also, of necessity, highlights the fluid character of genders, as well as their diversity. Hence, gender itself is also a moving target – an ongoing work-in-progress which, like the Information Society, we all participate in constructing.

We elaborate our analytical approach to gender and ICT further in Chapter 2 of the book, but already we can see how the broad approach is salient to our topic. It invites us to examine the dual, and mutually shaping, constructions of gender and the Information Society. And within this, it invites us to examine how gender(s) is/are perceived in relation to particular digital inclusion strategies – to unpack what stakeholders in such efforts mean when they invoke 'men' and 'women' as action categories, and to reflect on how these framings may shape particular inclusion practices. Digital inclusion strategies which address women are not only framed by conceptions of gender and of the Information Society; they are also framed by different strands of thinking about equal opportunities and how to work towards gender equality.

Equal opportunities: Category politics and gender mainstreaming

The language of equal opportunities for women emerged in the 1970s as a tool to tackle prejudice and discrimination and so build new standards of citizenship. Many countries established legislation to support these goals, which (amongst other things) put pressure on employers to change procedures and criteria for hiring and promoting. However, as Cockburn (1989) points out, experiences of equal opportunity practices led to disillusionment already in the 1980s. She claims that the failure of these practices was caused by insufficient activism from below, from trade unions and the grassroots. Equal opportunity had become too much of a bureaucratic fix, a top-down policy, according to Cockburn.

In a similar spirit, Carol Bacchi (1999) argues that the struggle for equal opportunities and affirmative action for women should be understood as an instance of 'category politics'. This concept highlights not only the need to develop policy instruments that address the problems encountered by recognisable social groups defined by gender, ethnicity, religious beliefs, etc., but also the need to analyse the construction of such categories and how they are deployed. Category politics with respect to gender means developing policies that provide positive action to improve the situation of women, working from presumed binary differences between 'men' and 'women' and identifying legal and other instruments to change the power relations underlying these differences. Clearly, to focus on women as a general category on which to base policy does not sit well with anti-essentialist feminist understandings of gender as fluid and diverse. Bacchi acknowledges this tension, but draws the conclusion that the use of category politics to support women is nevertheless useful and effective, quoting Patricia Williams (1991: 102): '[we need] to acknowledge the utility of such categories for certain purposes and the necessity of their breakdown on other occasions'.

Category politics is often performed as an intervention to rectify processes that disadvantage women, for example explicit or implicit discrimination or traditions of selective treatment. The main example is anti-discrimination legislation and the set-up of institutions to survey and fight discrimination and unequal treatment of women. To this end, Bacchi argues for the importance of affirmative action and reverse discrimination. She also seconds Cockburn's call for activism, emphasising that women need to engage in the politics surrounding the use of the category of 'women' (p. 164). Judy Wajcman's (2004) technofeminist position resonates with Bacchi's category politics in their shared call for feminist activism.

In the last decade, a different strategy has emerged, gaining widespread popularity in policy-making circles. It is called gender mainstreaming and is defined as:

> [A] gender equality strategy that aims to transform organisational processes by eliminating gender biases in existing routines, involving the regular actors in this transformation process [...]. Gender mainstreaming is regarded as an opportunity to counter the gender bias in regular policies, and to avoid the continuous reproduction of male norms in policy making (Benschop and Verloo 2006: 19).

Where category politics is concerned with the ways in which the category of 'women' is deployed in the politics of affirmative action and calls for women's activism to achieve gender equality, gender mainstreaming claims to work from the 'inside' and through policy instruments that are supposed to be integrated in the core activities of companies and institutions. Gender equality is supposed to be reached, not through dedicated committees overseeing the issue, but through routine decision-making. Equal opportunity legislation is less important to achieve gender mainstreaming than the use of instruments such as financial incentives, information regarding social and economic gains from diversity, and accounting devices to measure how women fare in the organisation. Thus, women's activism is less foregrounded.

From this perspective, gender mainstreaming appears as a kind of feminist reformism. As such, it is a contested concept (Walby 2005). Critics such as Stratigaki (2005) claim that the strategy has been shaped by goals other than those of gender equality, and that in practice it has been used largely as an alibi for neutralising positive actions emerging from category politics. Booth and Bennett (2002) lament the fuzziness of the concept, while Squires (2005) argues that gender mainstreaming holds some promise to be transformative, but that this depends on how it is interpreted. In any case, Benschop and Verloo (2006) conclude, the strategy demands a lot of work.

Our aim in this book is not to engage in critiques of either of these two approaches to feminist politics. Rather, we are interested in observing how each of them may actually be used as a basis for developing inclusion strategies. In principle, digital inclusion strategies may be framed in terms of category politics or gender mainstreaming, but possibly also by a mix of them. What kind of thinking dominates in any one case is an empirical issue. So we will ask questions like: To what extent are 'women' thought of as a distinct audience, target group or market segment, to be the object of particular action, in the spirit of category politics? And to what extent are actions trying to cater for the needs of women through 'products for everybody' or general political strategies, in the spirit of gender mainstreaming? How important is grassroots activism in shaping inclusion strategies, compared to top-down policy-making? And what are the consequences of either of these strategies in terms of the choice of instruments and actions?

The SIGIS effort

This book is the outcome of a European study entitled 'Strategies of Inclusion: Gender in the Information Society' (SIGIS). SIGIS is the largest single research effort undertaken to date on the subject of gender and ICT, and a very comprehensive one at that. The project took place between 2003 and 2005. It involved more than 20 researchers from five European countries, namely Ireland, Italy, the Netherlands, Norway and United Kingdom. It produced a total of 48 focused studies, an overview of which is provided at the end of the book. Altogether 30 different initiatives to include women into the design or use of ICT were investigated; a further 18 'user studies' solicited the experiences and meanings of selected strategies and ICTs for men and women who were targeted.

The inclusion strategies investigated were selected so as to provide diversity on a number of counts. First, they were initiated by a range of organisations: public sector, private sector and voluntary organisations. We were particularly interested to extend our gaze to market actors given prior indications that market actors might provide more promising inclusion initiatives than governmental ones. Secondly, in line with the ubiquitous character of ICT, the initiatives were targeted on people in diverse social activities and settings: education, workplaces, communities, homes and leisure spaces. Thirdly, they targeted diverse social groups, such as school children, teenagers, young computer enthusiasts, university students, residents in remote rural communities, residents in deprived urban neighbourhoods, teachers, civil servants, ICT specialists, unemployed mothers and young women.

Fourth, the inclusion strategies we studied cover a wide range of inclusion goals and measures, encompassing design – women *in* ICT occupations – and use – women *and* ICT products (cf. Faulkner 2000a). These include:

- Various measures to increase the recruitment of women into ICT occupations.
- Networks to improve the retention and progression of women in ICT.
- Basic ICT training for socially excluded groups.
- Vocational ICT training for specific occupations.
- Design of various new ICT products for women audiences.

Fifth, there are important differences in terms of how these strategies were framed. In most of the initiatives studied, increasing the inclusion of women was an explicit objective but in some it was a 'de facto' outcome. Some initiatives were designed specifically 'for women'; others were intended 'for everybody'.

Data for these focused studies were collected principally through interviews with relevant stakeholders and actors, supplemented where appropriate with written sources. In designing the research we considered what it meant to do an empirical analysis of inclusion, and came up with four main aspects:

1. What kind of mental models of the inclusion problem were brought by the actors initiating the inclusion activity?
2. Which set of activities or measures were seen as needed to achieve inclusion and thus constituting the 'action repertoire' of the initiative?
3. How was the strategy implemented?
4. What were the outcomes?

Following the above, our analysis of the data from the focused studies was concerned as much with processes as with outcomes, and with the conceptualisations shaping the strategic choices made by those doing the inclusion. This analysis has suggested a number of further distinctions: different kinds of inclusion thinking or processes underlying the different strategies. Particular features, and strengths and weaknesses, are addressed in the body of this book and drawn together in Chapter 10. It should be stressed at the outset that all of the inclusion strategies investigated were successful to some degree. Thus, gender inclusion into the Information Society can occur through a wide variety of strategies. There is no single magic formula.

Overview of the book

To recap, this book seeks to fill a major empirical and conceptual gap in the research literature on gender and ICT, by analysing strategies of gender inclusion in the Information Society. The path-breaking paradigm shift is, above all, expressed in the book's deliberate focus on mechanisms of inclusion rather than exclusion. The drivers behind this effort are twofold. From a policy perspective, the aim is to identify strategic features of different inclusion strategies in order to provide a better understanding of how and under what circumstances the gender digital gap may be closed. In practical terms, how might one implement slogans like 'An Information Society For All?' From a scholarly perspective, these inclusion initiatives provide very promising opportunities to analyse stabilities and instabilities in the performance or 'co-production' of gender and ICT. They are fascinating experiments in changing gender-technology relationships.

Another benefit of the inclusion paradigm is that the book necessarily foregrounds positive experiences with the use of ICTs, particularly among women. For example, several of the studies analysed in the book highlight the role of fun and glam as inclusion measures which focus on often invisible, undervalued and trivialised uses of computers. This is observed in relation to computer parties and computer toys and games, but also with respect to a much wider set of everyday life activities such as reading web-based magazines, chatting and using mobile phones. Last but not least, the benefit of an inclusion focus is that we get to see and analyse critically what works and doesn't work – as illustrated by the two examples we have just signalled above.

The book pursues this agenda through the next ten chapters. Chapter 2 outlines the theoretical points of departure and framing of the book. First we identify three narratives in previous research on gender and technology. The 'world without women' narrative is largely about the history of the active exclusion of women from scientific and technical fields, and about the consequences of this exclusion for the symbolic coding of technology as 'masculine'. With the end of overt discrimination and the general increase in the relative numbers of women in science and engineering in most Western countries over the last few decades, this narrative has largely been superseded. The 'chilly culture for women' narrative seeks to explain observations about many women choosing not to engage with ICT, and about a 'leaking pipeline' whereby proportionately more women than men leave specialist ICT work throughout the career trajectory. In relation to ICT more generally, there is a more optimistic counter narrative: cyberfeminism. This narrative has a clear focus on inclusion, but is criticised for reproducing gender stereotypes (the woman communicator).

We then argue that even the 'co-construction of gender and technology' work within feminist technology studies has focused more on technological change than gender change, often leaving gender binaries intact. In this book we analyse inclusion strategies through an invigorated 'co-production' framework, as shifting sociotechnical assemblages in which the performance of gender is addressed symmetrically alongside technology, as both hugely diverse and potentially fluid.

The remainder of the chapter explores what it might mean to think from inclusion, as opposed to thinking from exclusion. It signals, for example, a concern with what might motivate women to engage with ICT, and what might encourage them to stay engaged (as users or designers) – with socialisation as well as recruitment into the Information Society.

Picking up this challenge, Chapter 3 elaborates important aspects of processes of digital inclusion, addressing the most difficult nut to crack – how to reach the most digitally excluded groups. It draws on insights from research literature on the 'domestication' of technology whereby the appropriation and use of new ICTs is embedded in people's everyday activities, that is, home, work, friendship, education. We propose four key *inclusions needs*: access, capability, motivation and, underpinning the other three, various forms of support. We show that the most effective ICT training initiatives address all four of these needs.

We also highlight the importance of informal learning in digital inclusion, with people getting crucial help and encouragement from local ICT experts within their social networks. This vital feature of digital inclusion is often neglected in inclusion strategies, which in some cases may need to serve as local ICT experts for those they wish to reach. The point is illustrated through a particularly interesting community-based inclusion strategy, which was more successful in reaching digitally excluded women than men.

Chapter 4 examines a set of commercial web magazines and voluntary websites designed for women. The web magazines can be criticised for reproducing many of the stereotypes found in traditional women's magazines. Nevertheless, by recruiting and socialising large numbers of women into Internet use, these web-based inclusion strategies have achieved a symbolic re-construction in which women are increasingly seen as enthusiastic and competent Internet users – in contrast to earlier perceptions of the Internet as just one more 'toy for the boys'. We examine how this came about.

The web magazines and websites we studied attract women readers by providing content which is relevant to their lives, and by offering the motivation and opportunity to improve their capability in using the Internet. We call this phenomenon *self-inclusion*, or *assisted self-inclusion* where producers are proactive in meeting the inclusion needs of relevance and capability. In addition, women readers themselves became active in a third inclusion process, *socially embedded learning*, enhancing relevance and capability for themselves and other users through various forms of on-line interactivity. New content is produced (especially in the non-commercial sites), mutually supportive communities are built and women gain increasing confidence about using computers and the Internet.

Chapter 5 provides another example of progressive gender change, in this case through efforts to motivate children and young people's use of ICTs making them fun or glamorous. Prior research has assumed a very gendered toy/tool binary which recognises the playful engagement with computers by many boy and men enthusiasts but assumes that girls and women are only interested in using computers for 'serious' practical ends. Accordingly, strategies aimed at girls have neglected play and fun as routes to digital inclusion, as we demonstrate with a study of computing in secondary schools.

This toy/tool binary is challenged by the fact that both girls and boys have fun with computers in their leisure time. We investigate a computer party which brought together girls and boy computer enthusiasts. Whilst this highlights the social aspect of having fun with computers, participants' accounts of what they do reveal a gender dualised perception (which was not supported empirically) of boys as more computer skilled than girls. One reason for this may be a process of 'trivialisation' whereby it is the activity (downloading music, chatting, etc) rather than the computer which is visible to users. This certainly appeared to be happening at an initiative which sought to interest teenage girls in ICT by inviting them to design a fan website for their favourite pop band. Crucially, though, such strategies understand the importance of making ICTs 'instruments of desire'.

Somewhat unconventionally, the SIGIS project enquired whether ICT products emanating from the private sector can further gender inclusion in the Information Society. Chapter 6 addresses the 11 companies we studied whose ICT products (web magazines, games, and mobile phones) sought to reach girls or women. First, we identify a range of circumstances which may prompt such companies (mostly but not all small) to do this: a niche marketing strategy, the development of new technology, evidence of market demand and the influence of feminist entrepreneurs.

We then critically examine the four design methodologies these companies used, often in combination, in their efforts to be gender-sensitive. 'Relying on stereotypes' has the obvious danger of re-inscribing gender binaries, which may not be commercially rewarding. The 'I methodology', basing designs on what designers like, is problematic when the designers are very remote and different from the intended end users. A 'reflexive I methodology' seeks to overcome this disadvantage by involving (more) women in the design process, a strategy which presumes an essentialist understanding of gender. In principle, this recurring danger is avoided by various 'user participatory design techniques'.

The next two chapters shift the focus to the inclusion of women as designers and producers of ICTs and new media. Chapter 7 critically analyses inclusion measures employed to get more women into ICT as students of computer science, and to retain them. The literature on this has focused primarily on exclusion mechanisms highlighting various 'deficits' – in women, in computer science and in the image and culture of computing. Consequently, inclusion efforts are generally understood as needing to overcome these deficits, although the quite successful Carnegie Mellon's strategy appears to have built on both exclusion and inclusion perspectives.

This chapter investigates in a similar strategy. It used a variety of measures including advertising campaigns, a quota for women applicants, an information day, a computer lab dedicated to women students, and various forms of community building. Together these measures addressed all four inclusion needs, thus facilitating socialisation and retention as well as recruitment. They seem to be mutually reinforcing, creating a *positive circle of inclusion.* Nevertheless, we conclude that the use of quotas was critical, because increasing the relative number of women on the programme had a direct impact, on community building amongst the women students and on the visibility of women computer scientists. Significantly, the effort to challenge the masculine image of computing by mobilising a conventional feminine interest in people rather missed the mark.

Chapter 8 addresses an inclusion strategy which seeks to increase the poor retention and progression of women working in ICT, namely professional networks for women. The core argument here is that such networks can be effective in empowering women to progress their careers in ICT. Interviews with network members confirm the problems which make it difficult for women to 'belong' in a workplace or occupation dominated by men: that the workplace culture is more comfortable to more men than women, that the informal rules and contacts by which one progresses are passed on through 'boys' networks', and that the women tend to be invisible and stereotyped professionally.

The networks we studied help overcome these 'minority problems' in three ways. They provide a women-friendly community which the women belong to unequivocally and can gain strength from. Professional development is supported through various measures designed to increase relevant skills and confidence. And the networks seek to bring greater visibility and recognition to women in ICT. In addition, the very 'bottom up' character of these networks is empowering, and helps further the collective effort of members to improve the position of women in ICT more generally – in line with the emphasis on activism in category politics.

Some digital inclusion strategies are designed 'for everybody'; others are 'for women' only or chiefly. Chapter 9 compares and contrasts four rather different versions of the latter strategy: training courses, website and magazine, ICT products, and professional networks. We believe this important class of inclusion strategy may be characterised as involving the creation of 'women-centred spaces'. The chapter examines to what extent this approach can be an effective mechanism for furthering women's inclusion in the Information Society.

The women-only ICT training courses proved very successful in helping unemployed and disadvantaged women back into the labour market, thus confirming early second wave feminism's emphasis on women-only organisation as a stepping stone to equality. A similar conclusion can be drawn about the professional networks for women in ICT. In common with these two strategies, the on-line interactions of the websites and magazines reveal similar benefits of community building and mutual support around a common interest. We propose the term doubly articulated inclusion to capture the self-sustaining virtuous circle between content and community building going on here.

Cyberfeminism anticipates a 'feminisation' of the Information Society. In response, we emphasise that the target groups in all of these inclusion strategies are specific sub-groups of women, not women-in-general. And we wonder whether all of them might in time be superseded by forms of gender mainstreaming.

Chapter 10 analyses these and other policy lessons to be learnt from the inclusion initiatives investigated in the book. An 'anatomy of inclusion' is used to draw out the underlying strategic elements of these initiatives: the initiation (including what motivates public, private or voluntary sector initiatives); the underlying understanding of gender and of the inclusion problem(s); and the choice of instruments and actions.

Returning to our opening concerns about gender essentialisms, we note that stereotypes heavily shape the choice of instruments in many digital inclusion initiatives. We suggest that strategies which transcend gender may more be effective in the long run, especially in challenging the gender symbolism of ICT. For both pragmatic and political reasons, we encourage a politics of diversity, potentially combining category politics and gender mainstreaming. And we highlight the need to address motivation and relevance as critical in meeting the diversity of interests and needs which exists amongst 'women' (and 'men').

The chapter presents two main policy conclusions with respect to the development of inclusion measures: the importance of quantity and quality. The quantity argument draws our attention to the importance of increasing the relative numbers of women and men in ICT practices – as a means of overcoming minority problems through community building, and undermining the symbolic coding of ICT as 'masculine' by increasing the visibility of women using or making ICT. Numbers are at the heart of the way inclusion measures can work. The quality argument is based on observations that the quality of the measures can have a major impact on inclusion. Good quality products and delivery are more likely than poor quality ones to attract those currently excluded, and to keep them engaged.

Finally, in the spirit of encouraging shared learning amongst practitioners and of understanding processes of digital inclusion better, we propose that individual inclusion measures can be categorised by which inclusion needs they meet. We further propose the concept of *technologies of inclusion* to capture generic, potentially standardised ways of thinking about which measure(s) may be appropriate for any one inclusion goal, social group and setting. Seven of these emerge from the initiatives investigated in this book, all of which appear to work by the disciplinary power of a kind of 'normalising' which digital inclusion seems to require.

Finally, Chapter 11 presents some overarching theoretical conclusions. Whilst the defining focus on inclusion in this book implies a hopeful tack, our optimism about gender change in relation to ICT is tempered and ambivalent. The widespread use of gender stereotypes and binaries in the making of inclusion initiatives is worrying, given the tendency for discourses and symbolic representations to act as self-fulfilling prophesies. Yet at the same time, we observed real shifts in how gender is 'done' as a result of the inclusion initiatives we studied, albeit often modest. The new 'gender-ICT assemblages' so created have empowered many women, in relation not only to digital technology but also to other aspects of their wider lives.

Time and again we have stressed the diversity – of genders, of inclusion settings and of ICTs. The remainder of the chapter addresses our felt tension between arguing for gender diversity and claiming that gender does not matter. Gender needs to be considered, because inequality remains, but also in some important sense gender needs to be forgotten. When gender is emphasised too strongly, it is seen as more stable and dualised than it actually is, and this discourse may become self-fulfilling. We believe we have contributed to post-structuralist approaches by raising awareness of the complexity and multifacetedness of gender in relation to technology. (For example, our evidence that increasing relative numbers of women can serve to shift gender symbols suggest that symbols may be more fluid than is often assumed.) We invite readers to take up the challenge.

2

Changing perspectives on gender and technology: From exclusion to inclusion

Knut H. Sørensen

Gender and technology: What kind of problem?

In the previous chapter, we briefly introduced the research problems to be addressed in the book, emphasising the importance of conceptualising gender as complex, many-sided and fluid, rather than as an inert binary. Also, we argued the importance of approaching the topic of gender and technology as an issue of supporting inclusion as much as preventing exclusion of women from, in our case, the Information Society. In this chapter, we elaborate this argument by bringing in conceptual analytical tools.

As a social science concern, the gender and technology issue is basically a post-1980 phenomenon. It was largely the visions and outcomes of the large-scale introduction of computers into women's work that provoked the early feminist critique of technology, specifically gender, technology and work. New technology was observed to be, not a neutral force of production, but rather a tool that could be employed to undermine women's position in the labour force or as a basis for differential treatment of men and women (Cockburn 1983, 1985). The future of office work was prominent in many accounts of the emerging exclusion of women, due to a concern that women office workers, like secretaries, would be replaced by computer software that would be designed without sufficient appreciation of the work they did. There was also a concern that more men than women would be trained in high level computer skills, thus creating a gendered digital divide that seemed to give men unfair advantages. This tradition of research thus developed a topical interest in how the introduction of computers would shape and be shaped by the gender division of labour in workplaces (for an overview, see Webster 1996).

In 1983 came a pioneering report written by women students and research staff at MIT's Laboratory for Computer Science and Artificial Intelligence, describing the barriers they experienced as women in the computer profession. Their main complaint concerned differential treatment, in line with the research about women, computers and work. Sherry Turkle (1984) observed that the group of computer enthusiast students at MIT – so-called hackers – mainly consisted of young men who produced a culture Turkle claimed would alienate young women. In a similar vein, Sally Hacker (1989, 1990) concluded that the culture of engineering education was unwelcoming to women. Thus, a common argument emerged that the exclusion of women from the benefits of computers was the product not only of discrimination, but also of the institutions of computer education and research that developed a culture privileging men's (presumed) interests and abilities.

In the last 15–20 years, the approaches to studying women and computer technology have changed quite markedly. As noted above, such research efforts initially had a dominant focus on women, work and computers, which spawned a still outspoken concern about the comparatively low numbers of women in computer science and engineering (e.g., Cohoon and Aspray 2006a, Lagesen 2007, Fox et al. 2009). Then, as computers were transformed from computing machines to information and communication technologies (ICT), becoming household goods as well as workplace tools, there has been a growing interest in studying women as users of ICT in everyday life settings. Accordingly, two distinct problem areas emerged (see Faulkner 2001). First was a concern about women and ICTs, which spurred analysis of women's situation as users of the new technologies. Secondly, concern about women in ICT, which was pursued through research about women's situation as students and professionals of computer science and engineering.

In this book, we address both set of issues, analysing women and ICT as well as women in ICT. These dual concerns have been dealt with through a variety of concepts, such as digital divide (Norris 2001), underrepresentation (Cohoon and Aspray 2006b) or men's dominance (Wacjman 2004). The analysis of women in ICT has frequently been made into a separate issue (see the review of this literature in Chapter 7), and it has been noted that with respect to computing profession, there were relatively more women present in the early period of computers than it is today. In the collection edited by Thomas J. Misa (2010), the problem is formulated dramatically as "Why women are leaving computing". This differs from the digital divide approaches to women and ICT by claiming that the situation is going from bad to worse. When we still want to address both set of issues, it is because they appear to be related. First, similar mechanisms of exclusion and inclusion may be pertinent to women and ICT issues as well as women in ICT concerns. Secondly, a positive development with respect to women and ICT could also have a positive effect on women in ICT. Thirdly, more women in ICT might have a positive impact on women and ICT issues.

We frame this within a broader concern, which has resulted in a field of research usually labelled gender and technology. While this field is fairly diverse in terms of theoretical and methodological approaches, there is a number of efforts to review and synthesise an increasing number of contributions (e.g., Faulkner 2001, Wajcman 1994, 2004, 2010, Bray 2007) as well as readers and edited collections (e.g., Grint and Gill 1995, Creager et al. 2001, Yelland and Rubin 2002, Lerman et al. 2003, Fox et al. 2006, Schiebinger 2008). As these efforts show, the dominant concern has been the exclusion of women as users as well as professionals. While some contributions have looked into the men-technology relationship and the role of technology as constituent of modern masculinity (see, e.g., Hacker 1989, 1990, Mellström 1995, 2003, Turkle 1984), the exclusion of women has remained the major focus. This is even truer with respect to ICT (cf., e.g., Ahuja 2002, Archibald et al. 2005, Barker and Aspray 2006, Cronin and Roger 1999).

Given this fact and our explicit intention in this book to juxtapose exclusion and inclusion perspectives, we have chose to discuss previous research using 'exclusion' and 'inclusion' as organising principles. We start out by exploring what we consider to be the two main exclusion narratives relevant to women and/in ICT. The first may be called *'A world without women'* and focuses on absolute exclusion and the reproduction of ICT as a men's world. The second, *'A chilly culture'*, is a narrative about the problems of retaining women in a situation where they appear to be, and often are, in a minority as users or designers of ICT. In this way, we explore what understandings an exclusion focus may contribute to current concerns about women in the Information Society.

We then move to an inclusion narrative – that of cyberfeminism and the narrative of *'the woman communicator'*, which may be considered a counter narrative to the exclusion-focused ones. This narrative is concerned with inclusion mechanisms and provides an interesting point of departure for comparing the respective merits of exclusion and inclusion foci. However, from our perspective, the cyberfeminist narrative has clear limitations as an analytical as well as a practical tool. Thus, in the latter half of the chapter we try to elaborate what it might mean to think from inclusion, as opposed to thinking from exclusion, by engaging with the idea that gender and technology should be analysed as mutually shaped or co-constructed. Here, we review this proposal, which has become increasingly popular in the gender and technology field. In this way, we integrate theorising from the field of science and technology studies (STS), in particular related to the idiom of co-production (Jasanoff 2004), as a basis for the analysis in the rest of the book.

In short, the aim of this chapter is to examine inclusion and exclusion thinking about women and/in ICT as well as co-construction approaches, in order to provide conceptual tools for the following empirical chapters. We do not provide a comprehensive review of findings. However, throughout the book, the reader will find appropriate discussion of studies relevant to individual chapters.

Exclusion narrative I: A world without women?

Feminist scholars have for a long time argued that technology is very much a men's world. This narrative builds on a tale of long-term exclusion efforts on the part of particular groups of men. David Noble (1992) argues that these exclusion efforts were linked to the dominance of Christian clerics in science and the practices and institutions of higher learning during the last millennium. The ascetic culture of these clerics, says Noble, militated against the inclusion of women. Caroline Merchant's (1980) study of the Scientific Revolution in Britain also points to the reassertion of the Aristotelian binary associating men with rationality and women with nature, which justified the view that women were simply incapable of doing science. Ruth Oldenziel (1999) claims that the modern profession of engineering in the USA was also established through conscious efforts to keep women (as well as lower class or ethnic minority men) out, including symbolic representations of 'technology' and the emerging field of engineering as belonging to white middleclass men.

To put it bluntly, in this 'world without women' narrative, the historical outcome has been a culture of science and technology where women and femininity appear as matter out of place. What is produced and re-produced, according to the narrative, is an outspoken gender-based division of labour in relation to science and technology, combined with a gendered construction and appreciation of skills. It is claimed that this regime of divisions and differences is consequential in many circumstances, not the least with respect to women in science and technology. However, gender-related divisions and differences have also been observed with respect to use, from children's toys to men's and women's relationship to artefacts and activities in everyday life (Horowitz 2001, Lie and Sørensen 1996).

This 'world without women' is reproduced in part through a symbolic link between technology and masculinity, between boys and their toys (Lie 1998, Horowitz 2001), which makes it difficult to be a woman in the area because they are rendered marginal, numerically as well as symbolically. It becomes much more demanding to produce and reproduce an identity as a woman in a setting seen by others as overwhelmingly masculine and, thus, 'gender inauthentic' for women (Faulkner 2009b). ICT has been seen as subject to this patterning. For instance, some of the debates about computer games as mainly an activity for young men reflect explicitly the narrative of 'a world without women'. This illustrates a major difficulty with this narrative; it describes a massive exclusion effort (and outcome), which appears to be insurmountable.

Recent studies have instead tried to get inside the practices behind the symbols. They point out that some, mainly young, men's fascination with technology is embedded in their greater opportunity to play and work with gadgets and artefacts, and the advantage of being socialised into such behaviour (Faulkner 2000a, Kleif and Faulkner 2003). The roots of such fascination has been explained as an outcome of erotic relationships imbued with pleasures of exercising power (Hacker 1989, 1990); as a flight from the challenges of social relations especially with women (Turkle 1984); or as a way of reproducing a kind of brotherhood around technology (Mellström 1995).

To transcend the apparently monolithic narrative of 'a world without women', one could explore the actual dissimilarities in how different groups of men relate to technology. Of particular interest are tensions between the instrumentalism and rationality associated with science on the one hand and, on the other, the 'love affair' (Oldenziel 1999) with artefacts and tools. In Paul Willis' (1977) seminal study of the socialisation of boys into a British working class culture, the ability to cope with the strenuous physical relationship to machines – size, noise, need for strength, risk – was an important part of the construction of 'real' men, in contrast to the non-physical 'sissies'. Tinkering with technology often has physical aspects, and men professional engineers can struggle to combine intellectual and physical ways of relating to technology (Mellström 1995, see also McIlwee and Robinson 1992). The physical skills are important to their identity as men, while the intellectual approach is what sets them apart from non-professionals as well as non-engineers. Since computers are mainly intellectual machines, this balancing act may be less pronounced in relation to ICT.

Consequently, the narrative of 'a world without women' is not just a tale of the cultural foundations of the exclusion of women. It is also a story about how some groups of men get included into science and technology – not all men do – and how this inclusion happens. The importance of playfulness is often emphasised, as already noted, even though this aspect may be important to women as well (Nordli 2003). Indeed, research on women and technology may have overlooked the possibility that also women may be fascinated by technology, albeit perhaps in a different fashion (Berg 1996). Such observations are important if we are to avoid the 'a world without women' narrative becoming a self-fulfilling prophecy.

Also important in this connection is the (in)visibility of women's major role in the production of ICT devices and systems globally – an all too frequent omission even in much of the research literature. Women factory workers represent the majority of the workforce, particularly in the electronics industry in countries such as Malaysia, with low pay and bad working conditions (Rana 2008). Nevertheless, Ng and Mitter (2005) argue that ICTs overall has provided important opportunities for women in South-East Asia.

The 'world without women' narrative is largely a story about the history of the exclusion of women from scientific and technical fields, and about the consequences of this exclusion for the symbolic coding of technology as 'masculine'. This narrative has largely been superseded by an approach to the women and/in ICT issue that acknowledges the presence of (usually) more than a few women using or designing ICT following the general increase in the relative numbers of women in science and engineering in most Western countries over the last few decades. It builds on some of the same arguments about a masculine culture, but perceives exclusion in less absolute terms. Exclusion is definitively at work, but it may be countered. Thus, in the narrative of 'a chilly culture for women', there are at least some possibilities for reform.

Exclusion narrative II: A chilly ICT culture

Rossiter (1982, 1995) shows that historically the barriers to women who wanted to enter science, engineering and medicine included both formal rules stating women are not allowed and more subtle forms of discrimination (see also Berner 1996). In the face of the equal opportunity initiatives introduced in most Western countries in the last decades, most explicit discrimination seems to have been done away with. However, there remain cultural features of science and engineering in academic institutions and the workplace which facilitate more covert discrimination and differential treatment of men and women. Women continue to be a minority in science and engineering, even though their number has grown considerably (Hersh 2000, Fox et al. 2006, Phipps 2008, Schiebinger 2008). What are the implications of this minority status for women ICT professionals and users??

The idea of a chilly ICT culture is above all a response to the observations of women choosing not to engage with ICT, and of a phenomenon aptly coined the 'leaking pipeline' whereby proportionately more women than men leave specialist ICT work throughout the career trajectory (Millar and Jagger 2001). Accordingly, the term has been used in two ways. It has been employed to capture how women often experience ICT education and workplaces as unwelcoming – the more subtle practices which appear to flow from the minority status of women in ICT and which undermine their retention and progression in the field (see Chapters 7 and 8). The term has also been used to address the problem of recruitment: why many women appear reluctant to become involved with ICT. In this latter context, the chilly ICT culture narrative emphasises symbolic meanings of the technology – and so extends the focus to gender and ICT and to the practices of ICT users as well as designers.

Early on, computers seemed to be interpreted as a fairly gender-neutral technology (Sørensen and Berg 1987). During the 1990s, however, the initial promise of gender neutrality was seen to be corrupted because, increasingly, the use of computers appeared to be dominated by young men, while young women felt alienated (Rasmussen and Håpnes 1991, 2003). At the centre of this narrative lies the hacker figure. The hacker, who appears frequently in the literature, has in a way become the epitome of the chilling or excluding aspects of the culture of computing (Gansmo et al 2003a, b). Joseph Weizenbaum (1976) initiated the critical discussion of hackers' approach to computers, while Turkle (1984) added a gender dimension to the phenomenon. She described hackers as a contradictory phenomenon: they are admired because of their enthutiasm and skills, but at the same time pitied because of their lack of social competence and their social seclusion. And hackers are observed to be young men; no women on these premises. That hackers are overwhelmingly men is also evident in Levy's (1984) epic and celebratory expose of hacker history.

Later, hackers or nerds came to be cast more explicitly as the villains in the story of computer science as a chilly culture for women. Turkle (1988) found evidence of capable women opting out of computing because they saw the asocial aspect of the hackers as rendering computing, in Faulkner's terminology (2009b), a gender inauthentic option for themselves. Bente Rasmussen and Tove Håpnes (1991) found that nerds dominated the culture at the Norwegian computer science department they studied, and set standards in terms of what was perceived to be interesting and how one should engage with computers. The nerds were very involved with computers, something that the young women supposedly found appalling, in part because they were expected (as women) to want to engage in other activities than just sitting in front of a computer. A major problem, according to Rasmussen and Håpnes, was that the nerds made the women students feel second-rate and excluded from the centre of the discipline. Many of the women students lost interest in a discipline they perceived as shaped by and applauding the nerd image.

Jane Margolis and Allan Fisher (2002) make similar observations in their analysis of the situation at the Carnegie-Mellon University. They argue that if women are to feel at home as computer science students, the culture has to change. They see this culture as having developed to accommodate hackers' interests and ways of approaching the technology, singularly focused on some young men's 'need' to be in front of a computer screen as much as possible.

An additional problem occurs because student hackers have usually acquired their particular knowledge and skills during childhood and adolescence. This so-called 'boys' room competence' (Gansmo 2004) is an interesting phenomenon. It is widely assumed that the skills developed by boys using computers for games, programming and the like give them a head start when they begin to study computer science or engage in other ways of using computers. When many assume that these skills constitute the competence really needed in order to be an accomplished computing specialist, this represents a significant problem for young women (and some young men), who have not spent their adolescence in front of a screen and thus feel disadvantaged compared to the hackers when they start to study computer science. They are not, so to speak, up to the imagined standard.

It is important to note with Margolis and Fisher that the boys' room competence is not actually significant as a predictor of long-term achievement in computer science. Still, hackers' skills and screen stamina – combined with the apparent appreciation of these qualities by professors – do appear to alienate (some) young women from computer science. Even women computer enthusiasts appear to diminish their own abilities in relation to their men co-enthusiasts (Nordli 2003).

The problem reappears, perhaps not surprisingly, in the ICT industry. Woodfield's (2000) study of a UK software company started from the hypothesis that increased emphasis on communication as a core competence in computer science work should improve career possibilities for women. Initially, this assumption was confirmed by the stated recruitment policy of the company to emphasise communication skills and by their interest in recruiting women. Because women are generally assumed to be better communicators than men, women employees were expected to surpass men in the performance of 'hybrid jobs' where communication and ICT skills are combined. In practice, however, Woodfield found that the kind of general communication skills often attributed to women were not so highly valued. Rather, proficiency with computers and software were given more weight in the assessment of employees, by superiors as well as colleagues. Communication skills were assessed as a professional rather than as a personal quality or independent competence. Thus, the positive appraisal of many men's achievements was not offset by the effort to introduce new, more 'social' criteria in order to pave the way for more women. Arguably, these findings illustrate the dangers of mobilising gender stereotypes, and the fallacy inherent in thinking about communication as something distinct from the computer-related practices in which it is actually deeply embedded (Lagesen and Sørensen 2009).

The prevalence of the hacker or nerd culture in feminist accounts of ICT is surprising, given that several of its features seem to counter standard strategies of making computers useful to users and customers and, consequently, to making a profit. This is evident in the open source movement, which is attributed to a hacker ethos of providing (see, e.g., Castells 2001, Himanen 2001). From this perspective, hackers or computer nerds actually engage in very useful work and their love of autonomy help to open up new options in the choice of software. Consequently, they appear less other-worldly, perhaps also less exclusively 'masculine'.

The 'chilly culture' narrative which emerged in the 1990s, made the hacker figure into a strong force of exclusion and so came rather close to the story of a world without women. Still, it held greater optimism than the world without women narrative in its belief that the chilly culture could be transgressed. It opens the possibility that exclusion may be fought, for example, by critiquing the perception of men hackers as highly competent computer users and as a relevant role model for ICT specialists. This change in the understanding of what hackers or computer nerds represent, coincides with developments that have reduced gender differences in the use of computers, especially among young people, and made gender-ICT relations more dynamic. As we saw in Chapter 1, there remains a gender gap, but a large proportion of women – particularly young ones – have become eager users of ICTs.

In turning now to what we have presented as a counter-narrative – 'the woman communicator' – we seek to explore the effects of a shift to an analysis with a more optimistic focus on what might make computers attractive to women. Although this narrative does not actually display an explicit inclusion focus, it is nevertheless concerned with aspects of ICT use that might include rather than exclude women from the Information Society.

An inclusion narrative: The woman communicator

Some feminists have heralded the coming of the Internet as a positive turn for women's relationship with ICT. This belief – also called cyberfeminism by some (Bell and Kennedy 2000, Hawthorne and Klein 1999) – maintains that the Internet signifies increased emphasis on communication. Sadie Plant (1997) offers perhaps the most outspoken and radical cyberfeminist visions. She argues that women, computers, virtual reality and cyberspace are linked together in dispersed and distributed connections that she calls the matrix. She claims this matrix has an inherently feminine character, which will emerge to destroy patriarchy.

A more moderate and pragmatic version focuses more strictly on the 'communication turn' (Silverstone and Sørensen 2005), which denotes how ICT has come to represent a transgression of the idea that computers are made mainly for calculation and management of information. This is then assumed to be beneficial for women, since women supposedly find communication more interesting than calculation and will benefit from their comparatively better communication skills (Rasmussen and Håpnes 2003, Spender 1995). In her study of computer-interested young girls, Nordli (2001) found that they mainly used the computer for communication, information gathering and schoolwork. Similarly, the literature on ICT and women's activism (Harcourt 1999, Scott 2001) points to the communication and information gathering qualities of the Internet. Also Woodfield (2000) uses the presumption of women's communication skills as the basis of her initial hypothesis that women face improved career prospects in the ICT industry as performers of hybrid jobs.

Thus, cyberfeminist approaches share an optimism concerning women's computer-based activities, in particular the engagement with the Internet and net-based ways of communicating (Lagesen 2008). This optimism has been challenged, for example by Van Zoonen (2002), who criticises the idea that Internet is particularly suited for women. She claims that the Internet has been shaped by men, and that even women's everyday use of the Internet is frequently embedded in constructions made by men-dominated communities of designers. In a different vein, Adam (1997) criticises cyberfeminist positions for ignoring the massive empirical evidence documenting many women's negative experiences with ICT. Wajcman (2004) further takes cyberfeminism to task for leaning towards technological determinism in its emphasis on technological changes leading gender change. Instead, she argues, feminist political initiatives are needed to shape the Information Society in order to make it more welcoming for women.

Clearly, there are crucial weaknesses in cyberfeminist narratives, above all because they tend to reproduce an essentialist belief in 'woman, the great communicator'. Gender is thus portrayed as something constant; what is seen to change is the technology, in this case available computer applications. In this way, the main cyberfeminist argument invokes traditional dichotomous thinking about gender, but based on a reversal of the usual hierarchy. Here, what stereotypical women know (like communication) is deemed more important than what stereotypical men know (like programming). It may provide rhetorical support to include women in the Information Society, but probably at the cost of reproducing gender binaries. This needs to be explored.

A dynamic view of gender and ICT: The co-construction concept

All three sets of narratives outlined above provide important insights into the relationship between gender and ICT but also meet with serious limitations. The first was concerned with a strong, long-term socio-cultural momentum that has kept women out of science and technology, ICT included. Through this narrative, observations were made about men who do not want women in technology, who have struggled for a long time to keep women out, such that change appears difficult. The second was constructed from seeing how a minority of men has been able to impress their values and performances onto the practices and symbolisms surrounding computing, to the extent that doing computing appears unappealing or even appalling to women (and many men). The second narrative has largely replaced the first as the main story about the exclusion of women from ICT professions and advanced ICT use, without making change seem much more probable.

The third narrative, about the importance of communication to make ICT attractive to women and appreciating a skill stereotypically attributed to women, differs from the other two in so much as it is about change. However, all three are flawed by drawing upon a presumed binary opposition between the 'feminine' and the 'masculine' as an ordering device. Even cyberfeminist analyses treat gender as a constant while acknowledging technological changes. Thus, in a paradoxical way, much of the research about women and technology/women and ICT renders gender – or rather women – as immutable objects. It is technology and men that have to change.

This paradox becomes particularly conspicuous when viewed from the most popular idiom of feminist technology studies – the formula of gender and technology as co-constructed. The shift from talking about women and technology to gender and technology was a welcome move away from the tendency to view women as victims of men's technology (Berg 1994, 1996). In principle, the co-construction framework embraces and combines non-essentialist understandings of gender with constructivist understandings of technology, so providing an inherently dynamic perspective on both (Cockburn and Ormrod 1993; Cockburn and Fürst-Dilíc 1994, Faulkner 2001, Wacjman 2004). After all, what is construction if not making change?

However, as pointed out by Vivian A. Lagesen (2005) and later by Catharina Landström (2007), feminist technology studies tend in practice to treat gender and technology in an asymmetric way, with gender taken as a given point of departure. The object of study is gendered constructions of technology, whereas possible impacts of technological change on gender are rarely addressed. Thus, it is commonly argued that technology is central to predominant forms of masculinity (e.g., Wajcman 2004), while femininity may even be construed as a-technological (e.g., Turkle 1988). This in turn is used to explain why men become included and women excluded from technology.

Landström believes this asymmetry is caused by a dominant heteronormative perspective in the literature, which overlooks the way sexuality may work to diversify gender. Accordingly, she proposes the need to 'queer' feminist studies of technology. Lagesen argues that the failure to render gender as dynamic in the analysis is caused by a tendency to invoke gender as a structure, as a context of analysis. Consequently, she proposes to reform the co-construction approach by basing it more explicitly on action approaches, for example, by drawing on Sheila Jasanoff's (2004) proposed idiom of co-production – which departs from the fact that there actually is somebody doing gender as well as technology – or using actor network theory (e.g. Latour 1987, 2005) to analyse gender as socio-technical assemblages in order to see how the doing of ICT involves the doing of gender, and vice versa.

Other, action or practice-based approaches have actually contributed in a similar way. These approaches study in detail the actions involved in performing, for example, people's daily activities. One important example, which has been applied to the study of gender and technology, is domestication theory. This theory builds on (often) ethnographic studies of the complex processes by which people appropriate (or not) new technologies or new knowledge in their everyday lives (Silverstone and Hirsch 1992, Sørensen and Lie 1996, Berker et al. 2006). Although the term inclusion was not used in this literature, the domestication of ICT is an act of inclusion, a crucial feature of which being that it involves developing relevant ways of using ICT and of attaching meaning to the technology. ICTs become enacted through new practices and symbols. Crucially too, domestication studies not only provide insights into the use or non-use of ICT; they also demonstrate that new technologies may be used and construed in potentially quite diverse ways (Sørensen 2006). This diversity is central to our interest in change.

At the heart of such action or practice approaches is the understanding that stability with respect to performing, for example, gender should be analysed as an achievement and not be taken for granted. If gender appears to be stable, it is because it is made stable through the making and re-making of institutions, discourses, identities and representations (Jasanoff 2004) that uphold stable gender practices. Thus, to analyse gender and ICT we need to be concerned with how gender is done (West and Zimmerman 1987) and by whom – not just the making of ICT.

A key advantage of adopting a co-production framework is that, when we investigate the production of stability and repetition with respect to gender and ICT, we may at the same time observe sources of instability and change. By conceptualising gender as a heterogeneous socio-technical assemblage or the doing of gender as a socio-technical assembly, we may improve our understanding of gender dynamics in the stabilities and instabilities of these assemblages (Lagesen 2005). From this perspective, if a woman adopts a new mobile phone or a new piece of software, for example, she might end up doing gender in a different way than before, with potentially interesting consequences to investigate.

This book analyses a broad set of digital inclusion initiatives, drawing on concepts such as co-production, socio-technical assemblages and domestication. In this context, gender inclusion in the Information Society may be understood as a successful effort of co-producing gender and ICT, so that women as well as men engage actively with the technology without troublesome dissonance. Women should have access and opportunities to engage with ICT on par with men. Inclusion could also be understood as activities to provide motivation and resources to domesticate ICT.

The second exclusion narrative we described earlier is mainly centred on symbolic dissonances. It argues that women perceive and experience the ICT culture as chilly at least partly because of the way ICT has been represented as masculine, with men hackers as perhaps the main signifier of working with computers. This could of course also be read as a co-production, but one that stabilises gender divides along the lines observed in Chapter 1. Understood in this manner, the challenge in making inclusion initiatives is to reverse a co-production that results in symbolic dissonance between 'women' and 'technology', to undo existing representations. How may this be done?

Thinking from exclusion

The pattern of inequalities in ICT, or the digital divide, has concerned policymakers as well as academics for some time (see, e.g., Norris 2001, Castells 2001, ch. 9). A considerable body of literature covers many aspects of the problem with respect to women, and in diverse fields where the digital divide issue is articulated most pronounced, especially in education and training (Ahuja 2002, Dryburgh 2000, Cohoon and Asprey 2006b). As indicated in Chapter 1, the digital divide has usually been seen as a matter of access to computers and of having the skills needed to use them. However, Pippa Norris warns that we may need to reconsider the character of the problem:

> The explanation for the digital divide is often assumed to lie in certain characteristics of this technology, such as the need for computing skills and affordable online connections. The policy solutions designed to ameliorate the digital divide commonly focus on just these sorts of fixes, such as wiring schools and classrooms, training teachers, and providing community access in poorer neighbourhoods. Certainly this can do no harm. But will these

> initiatives work in terms of diversifying the online population? The results of this analysis suggest that, unfortunately, it seems unlikely. The policy fixes are too specific, the problem of social inequalities too endemic. … [T]he heart of the problem lies in broader patterns of social stratification that shape not just access to the virtual world, but also full participation in other common forms of information and communication technologies (Norris 2001:91–92).

Accordingly, Norris suggests that to overcome the digital divide – in our terms to do inclusion – one needs diverse and comprehensive strategies to meet the various inclusion needs. This seems in line with Manuel Castells' concern to ask critical questions about the implications of the digital divide:

> Is it really true that people and countries become excluded because they are disconnected from Internet-based networks? Or, rather, is it because they become dependent on economies and cultures in which they have little chance of finding their own path of material well-being and cultural identity? Under what conditions, and for what purposes, does inclusion/exclusion in/from Internet-based networks translate into better opportunities or greater inequalities? (Castells 2001:247–48).

Both Norris and Castells usefully remind us that the widespread concern about digital divides invites the use of simple fixes such as improved access to provide inclusion. Instead, they argue, we need to consider inclusion as a more complicated achievement (see also Tsatsou 2011). When analysing strategies of inclusion of women into the Information Society in this book, we therefore assess them not as instruments to persuade women to be interested in ICT per se, but rather as tools to support women to integrate ICT into their work and everyday life, to motivate and to help them make and get involved with new socio-technical assemblages (cf. also Ray 2004).

In contrast to this approach, it is common to understand inclusion as requiring the removal of the phenomena that produce exclusion. This is an understandable move, given that exclusion has been at the forefront of feminist studies of technology. It is a well-theorised phenomenon – for example, through co-construction analyses – and with a wealth of empirical underpinning. Moreover, it is well grounded in studies of social inequality, which in turn is a major concern of the social sciences, in particular sociology (see Liff 2002). That access to, use of and influence over ICT follows a quite standard pattern of social inequality (Norris 2001) is exactly what is to be expected. From this perspective, the main challenge is to combat inequality. This has been done in a variety of ways in the ICT area, often by trying to remove or change exclusion mechanisms.

So what are the main exclusion mechanisms identified in the literature – in addition to the symbolic dissonance discussed above? Dryburgh's (2000) review of efforts to explain why the number of girls and women in computer science has been declining in Western countries points to four main deficits or exclusion mechanisms:

- Structural factors including education policy, curriculum and organisation of computer science departments.
- Lack of role models and mentors.
- A masculine culture and the perseverance of sex stereotypes.
- Experience, attitudes and interest/motivation.

In a similar vein, Ahuja's (2002) review of research on the scarcity of women in ICT professions identifies a host of factors that influence career choice, career persistence and career advancement. Major influences include social expectations, work--family conflict, informal networks, occupational culture, lack of role models, problems with mentoring, and organisational/institutional structures. Robertson et al. (2001) argue that there are three processes in particular, which perpetuate the bias against women in ICT professions: (1) The marketing and pedagogic design of IT/IS/CS courses. (2) The exclusion of women from important and influential networks. (3) The difficulty in gaining legitimacy as an expert in fields such as IT and IS. In other studies we also find frequent mention of unfavourable working conditions, long hours of work, pay differences, unsupportive organisational cultures, and gender biases in the assessment of skills – together often leading to career disappointments (Amdam et al. 2005, Keelan 2008, Koput and Gutek 2010, Webster 2005).

In much of the literature, there is a tendency to view the broad field of science, engineering and technology (SET) and its practices as the main features to be changed, whereas gender – again – remains stable. While Cronin and Roger (1999) rightly point out that the predominant strategy to get more women into SET has traditionally focused on women's deficits, by trying to change women rather than SET, it may not be such a great step forward just to reverse the idea of what should be left constant and what needs to be changed.

According to Dryburgh (2000), the most commonly successful inclusion measure with respect to ICT and computer science education is based on some kind of women-only approach: particular classes or courses for women, computer labs only for women and the like. Educational strategy, the understanding of ICT and even of gender, may vary but the impact of 'women only' remains. Does this render men as the main exclusion mechanism? Henwood (1999 and 2000, see also Henwood et al. 2000) suggests that changing computer science course philosophy may not be all that important, because there seems to be a more fundamental force of exclusion – "some mechanism at work which continues to re-assert dualistic gender categories and identities in gender-technology relations, despite what would otherwise be very convincing evidence of the potential for their demise" (Henwood et al. 2000:127).

When one is thinking from exclusion – from observations of mechanisms that work to exclude women from making and using ICTs – inclusion efforts are as a rule constructed so as to counter exclusion. The main idea is to remove what appear to be obstacles. Thus, inclusion is approached as an epiphenomenon of exclusion, as a kind of conceptual side-kick that cannot be understood and analyse except in relation to exclusion. This is misleading, however, because it overlooks the crucial point that inclusion efforts must involve attractions: they need to start from an understanding of what it is that can make learning, using and working with ICT appealing to the digitally excluded.

In fact Webster (2005) and Amdam et al. (2005) do emphasise this two-sided quality in their analysis of women working in ICT. They observe exclusion mechanisms, but they also see that women value working with ICT and find it rewarding. Or, as expressed by Birnbaumer et al. (2007: 498):

> IT allows women to be independent in many ways: It is a challenging work, offering the opportunity for learning and perfection, and for being creative. It is also fascinating, satisfying the urge to 'solve riddles and generally the curiosity to get to the bottom of things'. It is an open world, with a wide horizon. It offers jobs, and this in a respectable area, and status. It helps the women to emancipate themselves from some of the constraints of their social and cultural environment.

Effective inclusion efforts may of course be constructed from exclusion thinking. Margolis and Fisher (2002) analysed a comprehensive and successful initiative to get more women to study computer science at one of the leading US universities in the field, Carnegie Mellon. These efforts to recruit and retain more women computer science students were based on a careful analysis of the exclusion mechanisms previously at work, and the main actions related to an image of computer science perceived as rather hostile by most women. Still, we think that the role of positive inclusion mechanisms in the construction of inclusion initiatives, and not just negative exclusion mechanisms, needs to be studied. Moreover, we suspect that a focus on processes of inclusion may actually provide a different and potentially complementary picture of the gender and ICT issue. How come?

Thinking from inclusion

There is a sizeable literature reporting on inclusion initiatives, although mainly those in education and training (Barker and Aspray 2006). However, these studies offer little theoretical or critical analysis – unlike, for example, Margolis and Fisher (2002). They tend to report from local actions or experiments, driven by very practical concerns. Moreover, as Alison Phipps (2006) observed, the initiatives themselves are often shaped by a desire to appear neutral and appealing to a liberalist understanding of equal opportunities, meaning an understanding that emphasises women's opportunities to choose differently than they traditionally have done with respect to education and training.

We might expect initiatives in the voluntary and private sector to differ from those in the public sector. In the voluntary sector, we find an increasing interest in using ICT as a tool for women and women's organisations to perform politics, as anticipated in the cyberfeminist literature. Some accounts emphasise the potential of the technology, in particular the Internet and related activities, to facilitate the performance of political work through easier access to information, and cheaper and quicker communication channels (Gibbs 1998, Harcourt 1999). Also, claims are made that local community feminist initiatives provide a more secure and productive learning environment for women to acquire basic skills in the utilisation of ICT (Green and Keeble 2001, Scott and Page 2001). Such initiatives actively seek digital inclusion on the basis of the relevance of ICT to potential users.

In the private sector, inclusion could be seen as coinciding with marketing, since marketing involves thinking from inclusion – how to positively motivate people to buy and use a product. Yet, there has been little interest in studying business initiatives from a gender and ICT perspective, with some exceptions (Oudshoorn et al. 2004). For example, Cassidy (2001) shows that, during the 1990s, the US computer industry tried to advertise the personal computer with a kind of feminine identity. These advertisements promoted the idea that personal computers were an important work tool as well as an instrument in developing family life. One of the most highly profiled initiatives to include women as users of ICT – in this case girls and computer games – has been the development of the 'Barbie Fashion Designer'. This game has been described as 'a beachhead in the dynamic dialogue between girls and computers'. It offers a glimpse into what Turkle calls 'the mirror of the machine'. "More importantly, however, it begins a new conversation" (Hamilton and Hokanson 1999). 'The mirror of the machine' in this respect highlights the possibilities in using ICT to foreground different and possibly evolving images of femininity to be experienced. The Barbie fashion designer game has also been heralded as an example of what Cassel and Jenkins (1998a:14f) calls entrepreneurial feminism, a strategy to promote feminist ideas by targeting new markets consisting of women.

Hendrik Spilker and Knut H. Sørensen (2000, 2002) analysed two examples of entrepreneurial feminism. The first was the design and marketing of a CD-ROM meant to help young women discover the world of personal computers and the Internet. They labelled this kind of inclusion effort women-in-particular because the design sought to attract young women specifically. The second was found in one media consortium's effort to make their web page attractive to all 'normal' users, women as well as men. This women-and-everybody-else strategy aimed to make women active users of the Internet by assuming that there is a standard, unisex mode for utilising the technology.

Spilker and Sørensen (2000, 2002) suggest that economic considerations may be the reason why some companies pursue a women-in-particular strategy, since there are many products that are marketed only for women. In addition, a segmented marketing strategy may be more effective than a general one. However, it may also be more costly. Thus we would expect the women-and-everybody-else strategy to be the more common. Spilker and Sørensen (2002) further suggested that private sector inclusion initiatives may in principle be more creative and radical than public strategies, since public authorities may have to be more politically and bureaucratically correct. It remains an open question as to whether this potential is realised in practice.

Thinking from inclusion means asking questions about why women might be interested in ICT (as opposed to why they are not). This approach highlights the importance of attending to relevance and motivation in digital inclusion efforts. Compared to an exclusion approach, an emphasis on inclusion therefore promises to provide better insights into what it is that makes (particular) ICTs relevant to (particular) people's lives, and what motivates them to engage with the technology.

In this book, we investigate whether this assumption holds, through the analysis of a large number of digital inclusion initiatives. We examine the performance of these efforts in order to investigate relationships between the thinking that underlies them, their concrete approaches to getting more women involved with ICT, and their outcomes.

Shifting from an exclusionary to an inclusionary perspective also means transcending the hitherto dominant concern with issues of access and recruitment into ICT. To become included into the Information Society one has not only to join but also to want to remain there (at least as long as one chooses). This condition points to the need to study inclusion in terms of socialisation as well as recruitment – to clarify the attractions and relevance of ICT on an ongoing basis, to catch the experience of being so-to-speak on the inside. In this book we consider what conditions have to be met in order to recruit and socialise people into the Information Society: motivation and relevance have already been noted, but are there other inclusion needs to be met? We address this question in the next chapter.

So, when we analyse inclusion strategies, we study what may be understood (abstractly) as conscious activities or sets of activities aiming to recruit people into and keep them within some system or practice. It is important to stress that we are concerned with the underlying model of the gender–ICT relationship as well as the strategic thinking about digital inclusion. For us, gender inclusion in the Information Society does not simply address issues of access, recruitment and socialisation; it also means looking at how gender and ICT are co-produced or re-assembled through inclusion initiatives. Our analysis of the gender and technology literature in this chapter has highlighted the need to develop digital inclusion measures that help transform gender as well as ICT. Important aspects of ICT cultures and practices may make inclusion difficult – as exclusion research anticipates – but there may also be facets of gender – gendered symbols and gendered practices – that need to be modified.

In this regard, we are particularly interested in the tendency noted by Henwood (1999, 2000) for dualised and essentialist understandings of gender to be reasserted (cf. Chapter 1). To become included in the Information Society has been interpreted as a kind of duty: everyone ought to know about computers, because this is useful and beneficial to society. In contrast to this instrumental attitude, the hacker culture has been described as a love of machines. Thus, in relation to narratives about ICT, love and duty have become gendered. Love, enthusiasm and play have been portrayed as characteristics of men's relationship with ICT, while duty and utility are seen as features of women's approach – and, thus, the most appropriate way to attract women users. We want to analyse how much inclusion strategies actually transcend such gender binaries, and with what consequences – for gender practices and symbols as well as ICT practices and symbols. In other words, what might a diversity politics of ICT look like, and how might more diverse gender–ICT assemblages be co-produced and nurtured? This is the core political project of our book.

3

Reaching for the digitally excluded

Wendy Faulkner and James Stewart

The focus on inclusion, rather than exclusion, processes with respect to technology is a defining feature of this book. We have argued that knowing about exclusion mechanisms does not tell us very much about inclusion mechanisms; the one is not a mirror image of the other. Rather, inclusion processes have dynamics of their own. Although there has been scant literature till our study on the processes surrounding digital inclusion strategies, we do nevertheless have rich insights into the general character of digital inclusion processes from the literature on the domestication of ICTs. Domestication studies have investigated in-depth how people do, or don't, go about adopting and using ICTs in their everyday lives. This research sought to shine light on the detailed nature of the processes which lie behind the diffusion of ICTs. In contrast to the heavy emphasis on market forces in diffusion studies (see Chapter 1), users and consumers of ICTs emerge as active agents in the diffusion process (Berker et al. 2006, Williams et al. 2005). The term domestication is intended to capture the sense in which users of a new technology 'make it our own' in the process (re)shaping it; technology is 'done' rather than 'is' (Lie and Sørensen 1996, Sørensen 2006).

People actively build new or altered practices and meanings as they appropriate new ICTs and incorporate them into their lives. These are key processes in the domestication of new technologies. We not only integrate the artefact into the routines and practices of everyday life, but also adjust or change our routines and practices around the technology and what it allows us to do. Crucially too, the domestication process has symbolic elements intertwined through these material processes. Technologies become 'objects of desire' – not simply or solely because we are persuaded that they can perform a useful function for us, but also because meanings 'attach' to the technology. Technologies acquire particular symbolic values or markers, which resonate (or not) with our particular identity, community and so forth. Early mobile phone use, for example, was seen to be the prerogative of 'important and busy people'.

So domestication involves complex but creative encounters between technologies and people through which non-adopters transform themselves into competent users (Silverstone et al. 1992, Berker et al. 2006). These encounters are of necessity heterogeneous; domestication involves sociotechnical changes which straddle social life and the technology. Thus, digital inclusion does not occur simply because people acquire or gain physical access to the technology. People have to have a reason to pick it up. ICTs must be seen to have some relevance to people's lives if they are going to make the effort to adopt the new technology. Relevance is a very situated thing: what is highly interesting for one person may be quite irrelevant for another. Many people have had no choice in the matter: they are obliged to learn about ICTs (computers especially) through work or education. But where there is a choice, making technology relevant to people's lives frequently involves (both the producer and other users) doing symbolic and identity work around the technology, to make it an 'object of desire'. As we will see in Chapter 5, having fun with computers, and joining in with peers who share these meanings, can be a powerful route to digital inclusion.

As well as needing a reason or motivation to pick up the new technology, people also clearly need to know how to use it. Following a long tradition in the study of technological expertise (e.g., Williams et al. 1998), we use the term capability here to indicate not only the skills required to use a range of ICT applications but also the capacity to make judgments, about how and when to use particular ICTs for particular purposes, plus the confidence to do so. Learning how to use ICTs effectively is a major and ongoing challenge, hence the understandable emphasis of many digital inclusion measures on training. Computer technology especially is plagued with technical complexity: confusing user interfaces, technical faults and technical limitations. Even when basic skills have been mastered, it can be deeply frustrating to discover that the technology does not always 'behave as it should'. The task of incorporating the technology into everyday activities is also troublesome for many users. And there is always more to learn because the technology does not stand still, with a myriad of new applications and versions to be coped with plus ever-changing expectations about what ICTs should be used for.

From these observations about the domestication process we can identify three key inclusion needs: access, motivation and capability. We view these as inclusion needs in the sense that they are all elements of digital inclusion processes which need to be present and addressed if inclusion outcomes – recruitment and socialisation into the Information Society – are to be achieved. For some people, these inclusion needs are already present; no intervention is needed to meet them. Thus, early adopters of home computers, who generally had the money, the motivation and the capability to learn by doing, needed to appropriate the new technology without much external support. By contrast, the so-called laggards are likely to be lacking on at least one of these fronts. Many simply cannot afford to buy ICT equipment, or do not currently see ICTs as 'objects of desire'. Some are not in work or education settings that oblige and support them to use computers; they may also lack the educational background, contacts and confidence to acquire the needed capability. It is in the reaching of these more digitally excluded groups that some kind of external support and/or intervention is likely to be most needed.

Whilst some of these inclusion needs are met through formal channels (Mick and Fournier 1998, Miles and Thomas 1996) – salespeople, user support services, training courses and the like – people's informal social networks are a crucial resource in the domestication of ICTs. Virtually everyone has one or more people in their existing social networks of family and friends who they turn to for encouragement and help in adopting and using ICTs: James Stewart calls these people 'local experts' (2002 and 2007), Maria Bakardijeva calls them 'warm experts' (2005; also Bakardjieva and Smith 2001). So, the adoption and use of ICTs are thoroughly embedded, not only in people's everyday lives but also in their social networks. For example, most people learnt to use the Internet from colleagues, family or friends (Katz and Aspden 1997, Wyatt et al. 2005). Indeed, much of what ICT users know about the technologies comes through informal learning – learning which is not certified or curriculum led but occurs, often spontaneously, as an incidental adjunct of everyday life and social networks (McGivney 1999, Livingstone 2001, Xiao and Carroll 2007).

The work of Stewart and of Bakardjieva indicates that local experts can play a key role in meeting all three of the inclusion needs identified above. In addition, they both found that much of what people need when struggling to adopt ICTs – and much of what local experts provide – is support. In this sense, the label 'expert' is too narrow since their contribution is not confined to providing information and building skills; they also encourage, enthuse and support. Some of this support is practical in nature: help with installing software or upgrading hardware, for instance, or by sharing or passing down equipment. But much of the support needed, and provided, is emotional and personal. Local experts help to overcome people's often deep feelings about the technology – ambivalence, frustration, even resistance (Otnes et al. 1997; Mick and Fournier 1998, Stewart 2003) – and they give people the confidence to 'have a go'. This leads us to view 'support' (of all types) as a fourth inclusion need, cutting across and underpinning the other three.

A final salient highlighted by the literature: by their very nature, domestication processes are local and situated. They unfold in a range of social settings or 'life spaces' – the home, workplaces, education, communities and leisure activities – and the social relationships and networks associated with them. Domestication studies have inevitably highlighted the particular, with their empirical focus on one social setting, group and/or technology. Early studies of ICT use in the home revealed how these various encounters feed into one another (e.g., Silverstone et al. 1992, Dholakia et al. 1996, Aune 1996); domestication processes cross the boundaries of our life spaces. This local and interlinked character of ICT adoption and use enables us to understand the specifics of digital inclusion processes – the particular things which motivate people (or not) to appropriate ICTs, the particular networks of ICT expertise available to them locally and so forth. As we will see repeatedly in this book, there is huge diversity in gender-ICT relations in any one setting, group or technology, and this diversity needs constantly to be borne in mind in the design of digital inclusion strategies.

This chapter explores these themes, with a particular focus on the challenge of reaching the most digitally excluded members of society. A central contention of our work on this is that local networks of ICT expertise are a vital element in digital inclusion processes, because local experts often play a critical role in meeting the four inclusion needs we have identified – of access, motivation, capability and support – where these are not in place. Yet the role played by local experts is often invisible to and neglected by policy makers, perhaps because it is informal.

We ask: What kinds of inclusion measures are required in order to meet the inclusion needs of digitally excluded groups? What about the role of local experts in relation to the more digitally excluded? And what specific characteristics of local networks of ICT expertise – including any gender specifics – need to be borne in mind?

The chapter is in two halves. In the first, we elaborate the main themes introduced here – local ICT experts and the roles they can play in meeting digital inclusion needs. We draw on examples from ICT training initiatives, since training is a central element of digital inclusion efforts, in the community and at work. In the second half of the chapter, we detail a community-based ICT initiative we are calling the Ardmore Network in a remote rural region of the Scottish Highlands. Typical of the enthusiastic rhetoric of the Information Society, the initiative sought to enhance social inclusion and community development, and reduce some of the disadvantages of remoteness, by facilitating greater use of ICTs. The initiative funded a network of local resource centres which, as well as providing ICT training, employed staff to promote and support ICT use. We have chosen to describe this strategy in some detail because it illustrates well a number of important conclusions about what is needed if inclusion strategies are to reach more digitally excluded groups.

Local experts and social networks

Social networks and personal influence have long been recognised as the dominant factors shaping whether and how people adopt innovations, technological or otherwise. Since most networks tend to link people with similar background and interests, innovations tend to 'trickle across' social groups rather than 'trickle down' (Rogers 1995). The more closed and narrow a person's social networks, the less likely they are to hear about or be encouraged to adopt a new technology. Key individuals facilitating the diffusion of a new technology are therefore often network bridgers who can translate and transfer ideas and knowledge across different domains of their life and different social groups (Bakardjieva 2005). The concepts of 'local experts', (Stewart 2002 and 2007) and of 'warm experts' (Bakardjieva 2005; see also Bakardjieva and Smith 2001) build on these ideas. We adopt Stewart's terminology here because his emphasis on the local, and on the way people are linked by networks, resonates with our concern about the specificities and situatedness of particular settings and groups.

In both Stewart's and Bakardjieva's research, virtually everyone they studied identified individuals they knew – either directly or through friend and family networks – who they regarded as 'experts' over ICT and who had supported them in various ways in using computer (and other) technology. These individuals did not have formal expertise in ICT – that is, expertise through specialist training or work roles – but they all know more about the technology than those who turned to them and/or were more confident about 'having a go' to work things out. Stewart found that some local experts are early adopters of new ICTs, enthusiasts who are confident all-rounders, often experimental in their approach to the technology. Others he characterised as accidental experts: they have the same enthusiasm for the technology and the willingness to 'have a go' as professional experts and early adopters, but they are 'ordinary people'. Perhaps for this reason, they are often particularly effective local experts; they can serve as very accessible role models.

Crucially, social networks are profoundly gendered – just as they are deeply stratified (in most countries) by divisions of class, race/ethnicity, age and so forth. These factors are likely to have a huge bearing on who does and does not have access to local ICT experts, and on who plays the role of local expert in any one case. Our research has revealed some predictable patterns – of local experts being men – but also some surprises. For example, amongst Bologna town council employees studied by Fortunati, some men put on training courses for colleagues [Fortunati 2003a: 344][13] whilst other men (managers) were reluctant to use computers (p.346) at all, and some women encouraged each other to attend computer courses or served as local experts to each other in Internet 'tearooms' [Fortunati 2003b: 393, 391]. Similarly, during a statutory programme of ICT training for Scottish school teachers, some women primary teachers turned to their husbands for further help on the computer, whilst others had women colleagues who helped and encouraged them [Faulkner and Stewart 2003]. Whilst professional ICT experts are more likely to be men, there is no a priori reason why this should be true of 'accidental' ICT experts. In many working class circles, for example, more women than men have encountered computers at work. The significance of gender in relation to informal ICT learning is not so much a question of whether local experts are more commonly women or men; the issue, as we will see in the case of the Ardmore Network, is how gender dynamics shape and are manifest in local networks of ICT expertise.

---

13 References in square brackets are to SIGIS reports, listed in the "References to the SIGIS focused studies" section (p. 273 f).

Roles of local experts in digital inclusion

Our research demonstrates how local or warm experts facilitate the adoption and use of ICTs by others in their social networks, in relations to all four of the inclusion needs we identified from the domestication literature. The first of these needs is access to the technology. Whilst many people have access to computer technology through work or education, and some make use of publicly available facilities, increasing numbers want to have access at home. Some people acquire their first computers as part of a 'trickle down' of gifting from earlier adopters or more advanced users – as happened with the handing down of mobile phones in the period of mass adoption (Sørensen 2005). Even in affluent countries, some sections of the population cannot afford to buy a computer, which is why public access technology is still needed. For those who can, purchasing a new computer can be fraught with difficulty. The uninitiated buyer risks misspending a considerable sum of money, whereas local experts often have the critical market and technical knowledge, so can make sensible judgements about what kind of system to go for and where to get the best deal. Through the stages of searching, choosing and setting up the new system, local experts can act as a bridge to professional and commercial service providers, mediating and interpreting these formal sources of ICT expertise.

The second set of inclusion needs concerns motivation. For those who are not introduced to ICTs in their work or education, inspiration from one's friends or family can be a key route to digital inclusion. Through casual conversations or enthusiastic demonstrations or in response to a specific enquiry – 'What is the Internet all about then?' – local experts inform members of their social network about ICTs and their uses. They help the 'idea' of an innovation diffuse into their communities, and so contribute to making it an 'object of desire'. Precisely because this person is known and local, they are able to identify specific uses or meanings of the technology that will resonate and inspire. Even just the knowledge that someone very close uses a particular technology for a particular purpose can have an important demonstration effect. Local experts can thus serve to promote change. A principal factor in the early diffusion of home computers, for example, was the evangelical work of experienced users, enthusing and encouraging new users within their social networks (Murdock et al. 1992; Rogers 1985). In such ways, local ICT experts help others to 'build practices and meanings' that make ICTs relevant or interesting in their lives.

The need for digital inclusion efforts to pay attention to motivation is nicely illustrated by a comparison of two publicly-funded basic ICT courses, one in the Netherlands [Rommes 2003a] and the other in Ireland [Hynes 2004]. Both courses were targeted on disadvantaged communities with high levels of unemployment and associated social problems, so both sought to reach the digitally excluded. And indeed, participants on both courses reported that their main reason for attending was a desire to keep up with and support their children's or grandchildren's use of computers and the Internet, and a desire to not be 'left out' of, or 'left behind' by, the Information Society. However, the Irish course appears to have been more successful in motivating participants than the Dutch one, in so much as a significantly higher proportion of students from the Irish course went on to become regular users of computers and the Internet, where some of Dutch graduates went on to become 'informed rejecters' (Wyatt 2004).

A critical factor here seems to have been that the Irish course started from the everyday uses which would be relevant and interesting to the course participants – such as designing household budgets on spreadsheet applications, or using the Internet to research possible summer holidays. The Irish course facilitated the domestication of computers and the Internet by showing trainees how ICTs might fit into their lives. By contrast, the Dutch course had a top down 'one size fits all' approach, which focused on basic skills – file management, finding web pages and so forth – rather than on uses. It failed to convey relevant meanings and uses to its participants (or to take account of the fact that many had no access to computers).

The third proposed inclusion need is the capability to use ICTs effectively. Even now that formal computer training is widely available in schools, universities and at work, informal learning through pre-existing networks of friends or family members frequently complements or supplements such training. This was confirmed time and again in our research. In the teacher training initiative in Scotland, Faulkner and Stewart [2003:287] found that the provision of formal training served to catalyse further informal learning, through colleagues and family members. The Bologna town council did not even provide training when it introduced computers; instead, it assumed its staff would learn to use a computer informally, in their homes or with colleagues [Fortunati 2003a]. By bringing material, training and assistance to the large majority of the population, local experts make up for the many gaps and limitations of the formal mechanisms for supporting ICT use.

The self-reliance and dedication of the local expert in increasing their ICT capability is a notable feature (Stewart 2007). Information searching and learning-by-doing are very time (and often money) consuming.

Little wonder many ICT users (and would-be users) are happy to lean on local experts they know; many are simply not skilled or interested enough to devote themselves to this. Many users opt to rely on local experts for acquiring ICT skills rather than formal learning or searching through manuals or just struggling with the computer – especially when a quick answer, or a slower, simpler explanation is needed. The convenience of being able to ring up a friend or ask a colleague when a problem arises, or of seeing someone else use a feature they may be thinking to acquire, is not trivial. Moreover, a local expert is often more willing and more able than a professional trainer to spend the time needed to help people acquire the more difficult elements of capability – the ability to make sensible judgements and the confidence to figure out things you don't know. By being similar to the user yet confident with computers themselves, local experts can serve as very positive role models.

The rapid pace of innovation in ICT means there is an ongoing and dynamic role for local experts in building up ICT capability locally. Because the technology is constantly changing, and its applications are ever broader, local experts tend to up-date their expertise continuously and that expertise becomes increasingly sought after by others. Networks of ICT expertise are not static. One informant in the Ardmore had (previously) played the role of local ICT expert in her neighbourhood, encouraging others to become competent ICT users and helping them with technical problems. Over time, this spawned new local experts: where no one in her street used to own a computer, there were soon three who did. 'The cover was always off' these computers as their owners exchanged parts or tested things out. Old computers were passed on to neighbours, who in turn came to these local experts if they had a problem (Faulkner and Kleif 2005). Thus, as one local expert's capability increases over time, so does the ICT capability (and use) of others in that person's social networks; there is a multiplier effect.

As will be clear by now, the fourth inclusion need – 'support' – is often crucial to meeting the three other inclusion needs. Many users rely on local experts for practical support with the things which exceed the boundaries of what they are able or willing to learn – the constant upgrades and additions (broadband internet, virus protection, etc.), functions changing and incompatibilities arising even in technologies that had apparently been 'tamed'. Practical support can include help with problem solving, copying software, passing on old or spare equipment, or pointing to an interesting website. Such practical support clearly underpins and facilitates access to ICTs, the motivation to use it and ICT capability building. In addition, Bakardijeva and Stewart both found that much of the support needed extends far beyond the narrowly practical or technical, to include social, personal and emotional support. This is the real merit of Bakardijeva's term 'warm' experts: these individuals are warm in the sense that they are known to and sympathetically supportive of the people who turn to them (Bakardjieva 2002). And Stewart (2003) concludes that emotional support is needed in the process of attempting to domesticate ICTs, because the technology can evoke such deeply unhappy or uncomfortable feelings – of being subjugated to technical or commercial forces, of being excluded or stupid due to lack of skills or resources, of finding it difficult to integrate ICTs 'meaningfully' into their lives.

Our research indicates that emotional support and empathy can facilitate both the motivation to adopt ICTs and the process of learning how to use them. This is likely to be especially important in relation to the more socially excluded and disadvantaged groups, who tend to know fewer people in their own networks who are experienced or advanced ICT users, and so may lack role models to inspire and help them to get going with computers. Also, socially marginal groups such as the unemployed typically have limited horizons and lack self-confidence, which can be a major barrier to 'having a go'. For this reason, good practice in ICT courses recognizes that building confidence and self-esteem – for instance, by celebrating all progress however modest – needs to go hand in hand with building ICT skills if the training is to be effective (e.g., Rommes et al. 2005). Broader forms of personal and social support are also often built into such ICT training initiatives, including support for child care, and a willingness to play the role of counsellors to participants whose lives may be harrowing at times (ibid).

It is clear that local ICT experts can play a very crucial role in facilitating the adoption and use of ICT – and hence digital inclusion – by helping people to meet the four inclusion needs we have identified. In our view, the importance of local ICT experts is as yet insufficiently acknowledged in the development of digital inclusion measures – a theme we will return to at the end of this chapter. Nevertheless a few caveats are in order. Not all experts meet the inclusion needs we have identified, and not all of them meet these needs equally well. Indeed, local experts are not necessarily better than professional experts: much of their knowledge may be weak or even wrong. Their close relationship with those they assist can create difficulties too – for example, if, power imbalances in personal relationships are exacerbated by the introduction of a new dependency.

Finally, not everyone wants to become an expert ICT user; they may be happy to rely on others. Some local experts take on a fairly permanent responsibility for ICTs in their communities or households – for example, to use a computer to purchase something or send e-mail, or to design web page. By providing such services and, by maintaining their expertise, local experts enable others to avoid adopting technologies themselves, as we will see in the case of Mharie's husband below. In such ways, many people who might be counted as non-adopters, in fact enjoy access to technologies by proxy. Policy interventions around closing the digital divide need to take cognisance of the role local experts play, not only in supporting less expert users, but also in enabling people to be 'digital citizens' without necessarily acquiring technical expertise and equipment themselves.

The Ardmore Network[14]

The Ardmore Network was a community-based ICT initiative launched with public money in 1997, 'to empower communities through access to communication and information'. It was part of a wider initiative which sought to help remote rural communities, by facilitating greater use of ICTs. Nine Community Resource Centres were established across the large but sparsely populated region – usually a room in a community centre. Each was equipped with various ICTs (a computer, software, printer, scanner, digital camera) and staffed part-week by a Community Support Worker. The Resource Centres offered three main services, to individual users and community groups: access to these ICTs and other informational resources; one-to-one support in using these resources; and ICT training. These services were promoted – through local advertising, word of mouth and outreach to specific groups – with the message: 'ICT is interesting, useful, fun and not scary; and the Resource Centres can help you get up to speed'.

In its very design, this heterogeneous package of measures had the potential to meet all of the inclusion needs described above. Our interviews with the staff and users of the Ardmore Network attest that this is indeed a good model for community-based digital inclusion strategies. It succeeded in increasing the level of ICT use in the communities they served, and so brought many real benefits to local residents: reducing the need for travel and generally 'making life easier' in remote communities through the use of on-line services (e.g., to purchase goods

14 Based on Faulkner and Kleif [2003], Kleif and Faulkner [2004].

or book holidays); being able to use technologies too expensive to buy (e.g., the digital camera); keeping in closer touch with distant family and friends; having fun and enhancing leisure pursuits; keeping up with the technology by gaining basic ICT skills; gaining labour market skills; and building self-confidence. In addition, two official evaluations found that the Ardmore Network had a significant impact in increasing ICT capability across the community, through the ICT training and one-to-one support. The size of the demand for training surprised even the Support Workers: 250 training places had been budgeted for in the first three years but in the event 750 were provided. Typically, these were short courses run over a series of evenings, with titles such as 'Computers for the Terrified'.

Some of the Network's visitors were already competent ICT users; they came to the Network's Centre to use ICT resources they did not have access to elsewhere. And some of those who came on the training courses were already motivated to learn more about ICTs; what they had lacked was the opportunity to do so. Advertising was all that was needed to reach them. But there was a third group of users for whom the simple provision and advertising of ICT training would not have been enough to recruit them; the computer reticent in the communities needed additional intervention to get them going. With this group especially, the Community Support Workers played two critical roles. First, they played a proselytising role, personally 'spreading the word' in their communities about computers, enthusing local people to start using ICTs and come to the courses. Second, they supported the ICT use and capability of many in their localities, through the one-to-one support they offered.

It is important to grasp that these Support Workers were selected for their community knowledge and commitment to community development, not for their ICT expertise, though they were all reasonably confident computer users. Because the initiative had a wider inclusion remit, they were charged with doing outreach work to specific groups, especially lone mothers and young people. One Support Worker described how she tended to work with any young mothers who were reticent about computers:

> They don't want a course. I wouldn't even call it training – that is too big a word sometimes. It is 'what would you like to do?', and going from that …

And another one put it in this way:

> Just tried to get chatting about anything that interested them and say 'why don't you come and have a look at the website? You don't have to do anything.' Just to try and start an interest off.

We see clearly the importance of finding relevance for ICTs. Similar exchanges took place with older women who visited the Centres. Dorothy and her friend Susan, both retired and members of a local art group, were persuaded by their local Community Support Worker to come on a basic computer course:

> He gave us the courage, because we said "Computers?! You must be joking! We're too old; we don't want to start again". [One of the first things he did was to] broaden our horizons; he "took us to art galleries" on the Internet. It just seemed so wonderful!

And here we see how confidence building can be a necessary element in this process. Dorothy went on to take three further courses and, with Susan, visits the local Resource Centre every Tuesday afternoon to use the computer there: as well as surfing the net, she emails her daughter in Sweden, shops on-line and uses PhotoDraw for her own artwork.

So the one-to-one support provided a means of motivating people about computers, as well as an opportunity for informal learning. This support was especially important in increasing ICT use and capability amongst the computer reticent or generally unconfident, who had often had bad experiences at school or around ICTs and felt uncomfortable about embarking on formal ICT training. For example, one woman had not gone near her computer for months because her husband once shouted at her for not shutting it down properly and told her she had broken it. One-to-one support works partly because it creates a safe, comfortable environment, and partly because the support can be tailored to the individual user – delivered to meet their particular interests and needs, and at a pace that suits them. One retired woman user met with the Support Worker over five weeks before she felt she could use Excel without help to do her household accounts. In the course of one-to-one sessions, the Support Workers frequently stressed their own lack of specialist ICT qualifications or work experience to users, but at the same time demonstrated, in their interactions with the computers, that this does not have to be a disadvantage. When they don't know how to do something they simply 'have a go' with the user present until they have worked it out together – a very positive demonstration of what an 'ordinary' user can achieve if they are confident enough to experiment.

The Ardmore Network strategy succeeded in reaching many digitally excluded people in the communities it served because its heterogeneous package of measures addressed all four digital inclusion needs – access, motivation, building capabilities and support. Several things were confirmed here. Learning skills was only part of what is needed in processes of digital inclusion: to get people to a point where they were willing to start using computers, the Support Workers had to find practices and meanings – its useful, fun, not scary – which were interesting and relevant to the potential users; and they had to build up their confidence around computers. Second, as we saw earlier in ICT training courses for disadvantaged people, the support provided needs to be about much more than the technology itself. The Community Support Workers brought commitment, sensitivity and local knowledge to the job. Their confidence with and enthusiasm about ICTs had a powerful role model effect. Because they were a familiar face in their communities, they could promote its use effectively and provide for those who might feel wary of computers a trusted bridge to the technology in a safe space.

In these ways, the Support Workers acted as local ICT experts for many in their localities. This conclusion is of great significance for digital inclusion strategies; it indicates that intermediaries can play the role of local expert for people who have access to little or no local ICT expertise in their own social networks (Stewart and Hyysalo 2008). By intermediaries, we mean any bodies located between the producers and consumers of a technology; this can include retailers, distributors, consultants, brokers and so forth, but also public sector agents such as the community support workers described here. The role of such intermediaries may be especially critical for communities and groups that do not have terribly wide networks or access to network bridgers – as is the case for many residents in remote rural or disadvantaged urban communities. After all, not being connected with a wide range of social networks is one measure of social exclusion (Granovetter 1973, Gilchrist 2004), which is likely to be linked to limited access to local ICT experts. The experience of the Ardmore Network suggests that intermediaries can make up for a dearth of available local experts and so reach the digitally excluded – if, and this is crucial, if they are effective in their outreach, in building familiarity and trust, in creating a safe environment, and in tailoring support and training to people's interests and needs as whole people. There are obvious parallels with the women-only training courses we studied since many on these courses made little mention of friends or family who helped them in their ICT learning (Rommes et al. 2005). The trainers on these courses can also be seen as intermediaries playing the role of local ICT experts.

How digital in/exclusion is gendered

The Ardmore Network's services were targeted on all members of the community irrespective of age, gender, employment or levels of computer literacy and expertise. They were promoted through local advertising and word of mouth, and through outreach to specific groups. There was a strong 'something for everyone' message throughout. Yet, the Ardmore Network strategy did not succeed in engaging all sections of the community equally. Every one of the Resource Centres attracted more women than men to the training and to the one-to-one support offered; largely absent, unless involved through community groups, were men of working age. Whilst the Ardmore Network strategy has successfully introduced previously non-using computer reticent women of various ages and social groups to ICT use, it failed to reach many non-using computer reticent men in the communities it served, especially middle-aged manual workers.

In order to find out why, we interviewed four women users and five men non-users of the Network's Centres (in addition to the Network staff) – chosen in an effort to capture at least some of the diversity present in the local communities. Adopting a modest technobiographic approach (Henwood et al. 2001), we asked three sets of questions: (1) How ICT literate are you and what ICTs do you routinely use? (2) How did you learn about ICTs and where do you go for help with ICT? (3) What use do you make of the Ardmore Network and why? The results are summarised in Tables 1 and 2. Even this cursory information from a handful of informants conveys the huge diversity which existed in terms of gender-ICT relations in the Ardmore region. In this section, we use this evidence to identify gender-related factors underlying digital in/exclusion in the region as well as the use/non-use of the Ardmore Network. The theme of local networks of ICT expertise emerges in both these regards.

Age and occupation were key determinants of ICT literacy and use in Ardmore as elsewhere. All of those under the age of 30 had acquired a basic ICT competence at school, whilst those over 30 who had learnt about computers had generally done so for work. Gender cuts across these factors, principally because employment and post-school education are quite gender differentiated in the Scottish Highlands. Most of the (local) men of working age in the region have permanent manual jobs in forestry, fisheries or fishing, where the women work mostly in seasonal jobs in tourism (bar work and cleaning), or in local offices. Women are more likely than men to work in jobs where computers are used, and more women than men go to university (partly because of the poor employment opportunities for women). Typically, the women who stay have limited horizons: they left school and had children early; they "don't believe in themselves much" and they "don't have any change left at the end of the week".

Table 1: Main findings for men who don't use the Ardmore Resource Centres

	ICT use and literacy?	How learnt?	Where go for help with ICT?	Use/non-use of the Ardmore Network?
Callum *Age*: 16 *Employment*: school student	*Literacy*: very good; "I've been around computers all my life" *Use*: graphics, games, searching the internet	informally: at home, from father formally: at school, later	strong network; friends come to him and he goes to his father	No – he has borrowed the digital camera
John *Age*: 26 *Employment*: manual work	*Literacy:* somewhat; "computers are not important in my life" *Use*: looks up things on the Net	formally: at school informally: from friends	many of his friends know "a lot" about computers one fixes their friends' computers	No – hasn't time
Jim *Age:* 32 *Employment*: unemployed	*Literacy*: none; I should learn but I'm "lazy"; I "don't understand the language" *Use*: none	dropped out of a basic computer course	many of his friends know "a lot" about computers	No – he uses the Community Centre for snooker [his friends would ask a friend, *not* the Resource Centre]
Davy *Age*: 45 *Employment*: manual work	*Literacy*: none; "I've no reason to use it" *Use*: none	n/a [too old to have learnt at school]	his wife is "very good"	No – once tried but left a basic course [possibly embarrassed to show ignorance]
Roy *Age*: 65+ *Employment:* retired	*Literacy*: good *Use*: email, searching for information on the internet	learnt at work	would buy in help if needed	No – uses the Resource Centre computer only when his own is broken

Table 2: Main findings for women who do use the Ardmore Resource Centres

	ICT literacy and use?	How learnt?	Where go for help with ICT?	Use/non-use of Ardmore Network?
Rosie *Age*: 16 *Employment*: unemployed	*Literacy*: good *Use*: email, internet, chatrooms (less frightening than meeting people face to face)	informally: "grew up with computers"; mother is an enthusiast formally: at school	strong network; part of a "big circle" of people who can help each other mother is a key local expert	Yes – because: she doesn't own a computer; there's a nice atmosphere; she goes with friends and has fun
Susan *Age*: 23 *Employment*: forestry	*Literacy*: good; can solve problems *Use*: email, internet, work; "I couldn't live without a computer!"	formally: early in school because dyslexic; free lap top for university	she is a local expert for her friends she and her mother are more interested than her father and brother	Yes – "it's my link out" (to friends met at university); her only access to email
Mharie *Age*: middle aged *Employment*: administration	*Literacy*: good *Use*: chiefly business administration applications	informally: on previous jobs formally: through the Ardmore Network	son is a computer enthusiast Ardmore Network Community Resource Centre if son is away	Yes – ECDL training
Dorothy *Age*: 62 *Employment*: retired	*Literacy*: good *Use*: email, internet, PhotoDraw	Ardmore Network: one-to-one support and basic training	Ardmore Network Community Resource Centre	Yes – support and training; now regularly uses the computer (can't afford her own)

This gender segregation of work and education points to greater digital exclusion overall amongst men than amongst women in the region, at least when the study was conducted in 2003. This was well illustrated by the contrast between the men partners of two women we spoke to. Susan was a university graduate in her 20s who "couldn't live without computers" because email was how she kept in touch with her university friends now she was living back home. By contrast, her boyfriend had basic computing skills but wasn't interested in using them; he didn't need computers in his work and had never left the locality so he didn't have distant friends. Mharie was middle-aged; she picked up ICT skills through office jobs in small businesses locally and was now doing the European Computer Driving License (ECDL) training. By contrast, her husband had not pursued formal ICT training even though he had some ICT skills and was out of work. She said he was unconfident about computers and "is not ready to admit that he doesn't know everything about it".

Three main groups of women made regular use of the Centres' training and one-to-one support, each with different motivations to increase their capability in ICT: young mothers, including lone parents, wanting to 'keep up with their children'; working age women wanting to improve their positions on the labour market; and retired-age women wanting to stay in touch with distant family or pursue hobbies and other interests using ICTs. By contrast, many men in the region appeared to have no interest in using the Ardmore Network to increase their ICT capability, again for a range of reasons. Some, such as Callum and Roy, were already very ICT-literate and had all the equipment they needed at home. Of those who did not need ICT skills for work, some used the Internet for practical purposes or leisure activities. Some, such as Susan's boyfriend, were simply not interested, and others did not have time to learn more about ICTs – either because of their work (e.g., fishermen spending long spells out at sea) or because of other interests (John did a voluntary job in his spare time). However, some men – like Jim, Davy and possibly Mharie's husband – were arguably motivated to learn more about ICT but seemed to be computer reticent. It seems likely that men whose identity was tied up with working in 'men's jobs', and who socialised principally with other men, may have been put off computers because they associated them with women's work and with education. It is also likely that many such men, like Mharie's husband, would tend to be reluctant to expose their ignorance to others.

Some of those who were ICT literate had become interested in computers through a close local ICT expert, rather than through work or education. Callum and Rosie both really enjoyed computers and got started under the early influence of computer enthusiast parents. (Both Rosie and Susan had computer enthusiast mothers.) Age, occupation and family all have a bearing on how much ICT expertise is available in one's social networks. Those who were young enough to have learned about ICT at school generally had at least some friends amongst their peers who owned a computer, including some who were very knowledgeable and enthusiastic users. The over-30s in Ardmore were less likely to have local experts amongst their (peer) friends; their local ICT expert if they had one was likely to be a son or daughter, who often no longer lived locally.

Arguably the most worrying group locally, in terms of inclusion in the Information Society, were those men and women over 30 years old who had not had the opportunity, at school or work, to learn about ICT and who had few or no local ICT experts amongst their peers or families. The greater success of the Ardmore Network in reaching women than men in this group must be explained in part by the fact that more women than men in the region were motivated to use ICTs. Still, there were men like Jim and Davy who were motivated to learn more about ICT but did not take advantage of the resources and support available in the Resource Centres. In order to explain this, we need to consider ways in which the inclusion mechanisms of the Ardmore Network strategy itself were gendered.

First, the marketing efforts of the Ardmore Network reached more women than men. These relied heavily on word of mouth which, as Harris and Dudley (2000) report in a study of Learning Centres, does not always reach socially excluded groups. To the degree that social networks in the region were strongly gender segregated and the Support Workers were all women, their efforts were bound to miss at least some men. Every Support Worker mentioned the importance of meeting people in the local shop, but not one mentioned the pub. Many told us they tried to reach the men through their wives, which in cases like Mharie was unlikely to succeed. And obviously, the outreach to youth groups and young mothers missed older men!

Second, the Support Workers could not be appropriate local experts for all groups in the community. At the time of our study, all but one of the Ardmore Support Workers were women and all but one were middle-aged mothers. There can be no doubt that the Support Workers provided a very positive role model for coming to use and learn about computers for many of the computer reticent women they worked with because they were in many respects 'ordinary women like them'. For those women who had limited horizons and self-esteem especially, this was someone they could identify with, much as we found with the trainers in the women-only ICT courses (see Chapter 9). It seems likely that the Support Workers simply could not serve as role models for many of the computer reticent men. We gained a strong sense that men such as Jim, Davy and Mharie's husband would not be comfortable exposing their ignorance about ICTs in front of others, perhaps especially a woman. Jim (32), who was unemployed and suffered from depression, would have liked to learn about computers, and knew the staff at his local Community Resource

Centre, but said he and his (mostly younger) friends would not visit the Centre;

> They can find out everything they want from their friends a lot of the time. They would go to their friends who know a lot about computers rather than go down there, you know? [...] I think it's because they can speak to their friends openly, in a way that they understand, really. Obviously, if you can go to your friends about some information about the computer, you would go to your friends rather than go down there.

As it happened, Jim chose not to expose his ignorance to his friends either, but our other studies of basic and vocational ICT training initiatives suggest that same sex role models work for both women [Faulkner and Kleif 2003] and men [McKeogh 2003a]. Gender, of course, intersects with other factors. Arguably, there is a case for having training and support provided by a more diverse range of staff – reflecting differences in age, circumstances and background as well as gender – though this may not always be practical.

Third and related to the second point, our evidence suggests that the Ardmore Network Resource Centres, like community centres more generally, were perceived and/or experienced as 'women's spaces' (see Laurie et al. 1999). The women users seemed to find the Centres comfortable places to visit, in ways that resonate with the women-friendly spaces described in Chapter 9. The Support Workers reported that retired age women and lone parents especially valued the social contact: it helped alleviate the particular isolation of these two groups. Sociality was also important for some younger people. Teenager Rosie provided a sharp contrast to Jim. Like him, she was not very socially confident, also unemployed and suffering from depression. But she visited the Centre because she found the atmosphere there friendly, and because she couldn't afford Internet access on her own computer. Like Dorothy, she brought her friends along to show them the amazing things they could do and find out about on the Internet.

Finally, adult education is not something men manual workers generally do. There is a longstanding pattern in community-based adult education whereby classes are dominated by women, especially in the over 40 age group, and the most under-represented group of men are manual workers and unemployed (McGivney 1998, ch. 1). McGivney (ibid., ch. 3) explains these missing men by a number of factors which resonate with the Ardmore Network evidence: a version of masculinity tied up with work and men peer groups, which carries anxiety about 'loosing face';

a perception that adult and continuing education is for women, partly because it is mainly staffed and attended by women; and a reluctance to admit learning needs (see also Tett 1994).

Our analysis of how the Ardmore Network strategy was gendered suggests the following practical implications: first, digital inclusion measures should look for ways to reach out to the digitally excluded in their existing social networks and spaces, as happened in the Ardmore Network with young mothers. One outreach option is to take adult education to the pub, for example (McGivney 1998, ch. 5, and 2000). Second, they should strive to create learning settings where the computer reticent might be more willing to expose their ignorance; McGivney found that a men-only course on ICT overcame the trainees' anxiety about using a computer in front of women who have better keyboard skills than them (McGivney 1998, ch. 5). Third, they should consider ways of making community network spaces more comfortable to all, and where possible use a mix of trainers and support workers reflecting the diversity of the community they serve.

One size does not fit all!

Our concern in this chapter has been how to reach the late adopters of ICTs, in particular computers. Many of the more digitally excluded are also socially excluded in one or more ways: they may lack the financial resources to acquire the new technology, have less formal education, less social mobility and status, and their social networks may be relatively narrow. We have argued that in attempting to reach the more digitally excluded groups, attention must be paid to all four sets of digital inclusion needs – not only access to the technology, but also the motivation and capability needed to use it effectively plus various kinds of support to underpin the meeting of these needs. This means that helping people to gain the benefits of using ICTs requires considerably more than simply providing access and training. Various forms of support – practical, emotional, personal and social – are especially critical in reaching the more digitally and socially excluded sections of the population – both to get people interested and motivated to get going with computers and to underpin their learning. One obvious conclusion of this is that ICT training measures must move beyond simple skills acquisition: it must recognise both the particular drivers that are likely to motivate specific groups to pick up ICTs, and the various kinds of encouragement and support needed to be effective in promoting greater use of ICTs and building the necessary capability.

We have shown that local experts are a crucial ingredient in the diffusion and domestication of ICTs – and hence in digital inclusion processes. They are able to help meet all four inclusion needs identified in this chapter, in part through their enthusiasm for the technology and their sustained efforts at 'learning by doing', and in part by their very availability and their familiarity with those who seek their help. This is something that has remained largely invisible or un-remarked-on in policy circles. Moreover, we have shown how staff in a community-based ICT initiative can play the role of local experts.

An obvious, and we believe very valuable, conclusion of our research is that intermediaries – like the Ardmore Network Support Workers, and some of the trainers we have seen in ICT courses for disadvantaged groups – may be needed to facilitate digital inclusion for groups and communities where local networks of ICT expertise are weak. This has real policy implications, given the dynamic role of local experts in the diffusion of ICTs. Where people have little or no access to local ICT experts in their own social networks (often linked to social and digital exclusion), interventions are needed to create intermediaries who can act as local experts and so create more ICT experts locally. It may also be important to locate and nurture any local experts that already exist in the community in order to facilitate ICT use and (informal) ICT learning, especially amongst social groups and networks where this is rare – such as amongst middle-aged men in manual jobs.

Finally, in Chapters 1 and 2, we highlighted the problem of using the categories 'women' or 'men' as a point of departure for digital inclusion strategies, not least because of huge differences amongst women and amongst men. Because social networks and gender are both thoroughly embedded in people's everyday lives and relationships, local networks of ICT expertise are themselves bound to be gendered – in a variety of complex ways. An obvious corollary of this diversity, confirmed time and again in our research, is that 'one size does not fit all' with respect to digital inclusion strategies. The Ardmore Network proved to be a particularly stark illustration of this: it demonstrates very clearly not only that digital exclusion mechanisms are gendered, but also that digital inclusion mechanisms may themselves be gendered – in very particular ways. This has relevance right across the range of strategies we studied. It means that digital inclusion measures must be effectively tailored if they are to reach all sections of the population. And effective tailoring means being aware of the particular social locations, identities and so forth of all the diverse groups targeted. Specifics matter!

This is a real challenge to 'for everybody' inclusion strategies. It means that practitioners really must address the diversity of their target groups, and identify likely gender and other dynamics present, if they are to reach everybody. It also means that practitioners need to consider the particular ways in which processes of digital inclusion might be gendered, if they are to avoid missing significant (and identifiable) sections of the community. The point may seem self-evident, but it bears stressing in a context where digital inclusion initiatives intended for both women and men in a particular group are frequently 'gender blind' – that is, they fail to 'see' and take account of any gender dynamics surrounding digital in/exclusion mechanisms amongst the target group – or where any 'gender awareness' is usually interpreted as looking for presumed differences between (all) women and (all) men.

4

Websites for women: New routes to digital inclusion

Els Rommes

The previous chapter looked at inclusion processes in some public sector digital inclusion strategies, and analysed them in terms of what inclusion needs have to be met if the strategies are to succeed. In this chapter, we look at a rather particular set of private sector and voluntary sector initiatives – websites targeted wholly or mostly on women – and analyse these in terms of the inclusion work required. Because websites are, even in our Web 2.0 age, a central (socio)technology of the Information Society, our primary concern is to understand what particular opportunities and challenges they bring as a potential instrument of digital inclusion. In Chapter 1, we saw that women now constitute an increasing proportion of Internet users, and here we see that various kinds of websites in the past have contributed to that greater ICT use by women nowadays. So we ask: Are we seeing new routes to digital inclusion through the use of (interactive) websites? And what kinds of inclusion processes are involved?

We propose the term inclusion work to capture any activities – on the part of both users and producers of ICT – which contribute to furthering users' inclusion in the Information Society by meeting one or more of their inclusion needs. This formulation means that we are interested both in the nature of the work and in who is performing it. As we argued in Chapters 2 and 3, digital inclusion is a process of domestication. Inclusion work, therefore, involves the creation of new practices and meanings, which contribute to domesticating the ICTs, to 'making them our own'. So, as well as asking what inclusion needs are being met by the inclusion work – access, motivation, capability and support – we are also interested to see what new practices and meanings emerge around technologies. How has the Internet been domesticated as a place where many women now 'feel at home'? What have been the processes of recruitment and socialisation? What transformations in either gender or the ICTs have occurred as a result of the development and use of these websites? What kind of symbolic work and transformations has accompanied the growth in the numbers of women using the Internet? And to what extent have these replicated or challenged gender dichotomies?

Inclusion work can be performed by producers of an ICT and/or by its users. In general, we would expect that the more inclusion work is done by producers of a ICT the less needs to be done by users, and vice versa. Moreover, the more inclusion work producers do, the more trivialised the technology becomes to the users. By trivialised we mean that the technology is rendered a matter of course to those using it. It is not necessary to understand how the technology operates to apply it (Weingart 1989). It is after all the relevance of the technology, what it may help to achieve, rather than the technology per se which attracts users, thus 'normalising' the technology as part of everyday life. We are interested to see how the respective roles of producers and users in the inclusion process play out in terms of the different inclusions needs, and what the implications are in terms of inclusion outcomes. Is there an optimum model here?

To address these themes and questions, we draw on studies of six websites. These comprise four commercial women's web magazines (the Norwegian web magazine 'Femme', the Irish 'iVenus', the Dutch 'Libelle' and Italian 'Donna Moderna'); and two voluntary websites – the Dutch site Women On the Web ('WOW') and the Italian 'Sensa Maschera'. All of these had an early web presence: they came on-line between 1996 and 2000 and so have influenced Internet use from an early phase on. All of the websites except 'Femme' were still on-line in 2011 and had not changed much in their looks or functionalities after we studied them. Moreover, over a decade after their conception and despite the availability of alternative social network sites like Facebook or LinkedIn, we could still observe lively and large communities at these websites. The number and content of the contributions by users to the interactive parts of the websites were proof of that.

The studies we drew on for our analysis were based on thorough examination of the websites as well as interviews with a cross-section of the staff responsible for design and production them. In addition, users of Sensa Maschera, Donna Moderna, WOW and iVenus were studied; the first two by a survey questionnaire and the latter two by on-line interviews.

The first section of this chapter describes the four commercial women's web magazines in some detail, so as to get inside the inclusion work involved, and reveal the kinds of practices and meanings which were built as these websites were domesticated. In the second section, we propose the term 'assisted self-inclusion' to capture the combination of inclusion work done by producers and users. We found an interesting contrast between the commercial web magazines we studied and the non-commercial websites in this regard. In all cases, however, the relevance of the content is clearly a central element attracting potential users to the site. The next section investigates a second form of digital inclusion which emerged – inclusion through on-line interactivity – which proved to be a powerful mechanism for meeting inclusion needs, especially for supporting ICT capability building, in all six websites. We then discuss the symbolic transformations which have accompanied both forms of inclusion, highlighting in particular that the Internet has become for many women a place where they 'feel at home'. We conclude by drawing out some general themes that might be applied to other kinds of digital inclusion strategies.

Commercial women's magazines

Femme (Norway)[15]

Femme was a commercial web-based magazine, aimed at women in Norway. Launched in June 2000, it was originally only available on the web so did not have a printed version through which readers could be directed to the website. The initiators also started producing a print magazine one and a half year later when the web magazine struggled to raise sufficient advertising revenue. In the end, however, these funds were not enough and Femme filed for bankruptcy.

15 This section is based on Hestflått [2003].

The editors and producers of Femme had a clear picture of the kind of women they hoped to attract – namely, "women likely to use the net", who they saw as "young, single and urban", with a relatively high income and education – in short, "women like ourselves". Their image of their readers was also later shaped by the kind of women who they discovered did use the magazine. As the editor of Femme commented:

> We are not for every woman between 10 and 80 years old, to put it that way. Our users are mainly between 20/25 to 35 years old. Our target group changes when we see whom we actually meet. For example, we know that to a large extent our readers come from the largest cities. Either they are full time employed or students, close to 50-50, and 53% are single [laughter], which is more than the average, which is 40%. Most of them have higher education, and they earn [a middle income].

The Femme staff perceived their readers as women who want to be 'different from the rest', who have a sense of humour and self-confidence, who 'make it on their own' or who present themselves as 'one of the boys'. They saw these women as having a critical attitude towards feminine stereotypes like 'skinny models' and the 'typical housewife', so they chose to have a chubby famous Norwegian stand-up comedian as their 'brand' woman – a 'Miss Femme' who did indeed look like someone who goes her own way, who is strong-willed and slightly ironic, but at the same time has a feminine appearance, with long hair, body curves and large eyes.

There is a clear tension between the Femme producers' image of their target audience, which sought to avoid stereotypical or normative versions of femininity, and their design choices regarding content, lay-out and navigation, which were borrowed heavily from existing print women's magazines and other web magazines for women. The editor again:

> A lot of things are picked straight from *Cosmo*, like the organisation of material. It goes without saying that you cannot start a wide-reaching magazine without things like fashion and beauty.

Content was chosen, according to a designer, on the basis that "certain things just belong to being a woman", and so certain topics "just belong on a women's website" – beauty and fashion, sex and love, health and well-being, and home and family. Moreover, essentialist views on gender were often expressed in editorial comments on the website, along the lines that men and women 'just are different'.

Essentialist notions were also evident in Femme's lay-out and navigational choices. One of the web designers remarked that their lay-out contained "many visual cues that make women feel comfortable with the design: colours, photos, words". They chose to use "feminine colours: pink and violet", which they balanced "with some blue, so that the result altogether does not end up very feminine". The priority in the design of the navigation of the website was to make it 'user friendly':

> The user interface is important. Femme was, for example, built in such a way that it should be easy to understand, to use, to read. Big headings, links easy to see. It should also make a visual impression, so we use many pictures – we worked a lot with those things in the beginning. So that it should meet the tastes of the target group.

It seems the designers assumed that women might have problems reading, understanding and navigating websites, and they wanted to "educate their users gradually".

So where had the original ideas of the editors and designers of Femme about their target group as being 'different' from feminine stereotypes gone? There were content choices in which the originally intended target group could still be recognised. Several topics directly related to their younger, higher educated, higher paid target group, such as, 'Travel and Adventures' and 'Career and Money'. One topic, 'Cars and motorcycles' was surprising even for a magazine aimed at 'modern women'. The editors had decided to include this topic in part for commercial reasons, but also because "nobody else does it" and "we want to educate, to do some nurturing". Again mobilising essentialist notions, they felt that, because "women stress different things than men", they would want to read different things about cars than men.

Libelle (Netherlands)[16]

With 2.3 million subscribers in 2001, the women's magazine, Libelle, is one of the largest Dutch paper magazines and the brand 'Libelle' is well known in the Netherlands. Compared to the other web magazines, it is quite traditional: its target group consists of middle-class, heterosexual women who are 35–55 years old, middle to highly educated, who work part-time or are full-time housewives and who have (young) children at home. The decision to launch the website was a bold one in February

16 This section is based on van Slooten [2003].

1998, given that this target group was not generally associated with computer and Internet use. The website was promoted in the paper magazine, on various websites and during the popular Libelle Summer Week (where women can follow workshops on the beach), making it one of the most popular 'for women' places on the Internet at that time.

One of the main reasons for launching the website was commercial. The market for mass magazines had been decreasing, and so the publisher of the magazine, sought new ways of making the brand 'Libelle' attractive. It was hoped that publishing the magazine on the Internet would give Libelle a younger, more trendy and modern look, and encourage additional advertisement in both the web and paper magazines. In the event, the web magazine did indeed attract somewhat younger readers, about half of whom did not already read the paper magazine. In the event, the website proved to be a useful vehicle for the editors to find out more about the preferences of their readers.

As the producers of the paper magazine sought to attract more or less the same audience with the web magazine, they mostly copied the 'Libelle style' and topics like cooking and leisure activities. To this, they added a very popular weekly puzzle, the option to send e-cards, an electronic newsletter, web TV and information on the use of computers and the Internet. They wanted their website to be accessible for people with old computers and slow Internet connections, so they used pictures that were not too heavy to download.

Above all, the designer of the Libelle web magazine felt passionately about ease-of-use, insisting that navigation should be "super, super simple; always visible, always knowing where you are, always recognisable." As the art director commented:

> If you design a site for hackers, for example, it would be a very different design than a website for the 'Libelle reader'. No, I think that you need to keep a site like that for Libelle just really clear and give the users their information without too much hassle.

Many features made the website 'easy to use': the frames were fixed with clear titles to show that you were still on the Libelle website; the colours of the site would change depending on the topic; symbols were used and a large letter type; and not too many options were offered on any one page. In addition, many features sought to increase users' skills in using the website and computers more generally: more complicated parts of the website, like a web-TV option would guide the reader step-by-step through the choices they had to make; there was a place on the website where users could practice their mouse-skills; a 'computer alphabet' provided a translation of the most used computer terms and jargon; 'tips and tricks of computers' were added to the electronic newsletter of Libelle; readers were invited to attend workshops to learn how to use computers and the Internet during the Libelle Summer Week; and 'technical' questions could (initially) be addressed to the webmaster. Later on a heavily used interactive computer forum on the website was installed so that the webmaster did not have to spend all her time answering questions such as 'How can I download the Internet' or 'How can I block unwanted e-mail'. Clearly, the makers of the website assumed that the target group of Libelle did not have many computer and Internet skills, but were interested in acquiring them. By 2007, it was assumed that women now have these skills, as virtually all the technical support facilities had disappeared from the site.

Facilitating the use of the Libelle website in these many ways had an obvious commercial rationale: if they wanted their target group to visit the website, they needed to make it easy to access and provide the necessary skills. But this does not entirely explain why they went to such lengths to increase readers' more general computer and Internet skills – which made it a more general inclusion strategy than found at the other websites. Significantly, Libelle facilitated women in acquiring broader ICT skills because their users asked for this and indicated that they were interested in computer topics. According to the editor-in-chief, Libelle deals with topics that are 'hot' in Dutch society "and today that is politics, but also Internet". However, users might not have asked for these options if Libelle was not provided them from the first issue. So the choice to include computer items on the first website design was probably crucial in making the Libelle website a more general inclusion strategy. The designer of this first issue had an explicit inclusion agenda: she wanted to "reduce the technical barriers for women" and "convince [them] that it [computers] was not a difficult world to participate in".

iVenus (Ireland)[17]

iVenus was set up in 1999 by a company which publishes three of the four main Irish women's print magazines. Its main purpose was to establish a web presence for the firm and the content it can produce, which would roll out across Europe and become the main portal for the English-speaking world. Accordingly, a huge investment was made in the tech-

17 This section is based on MacKeogh [2003e] and MacKeogh and O'Connor [2004].

nology, so that it would have the potential to be accessed multi-lingually and would provide a range of interactive services. A country-wide promotional campaign was launched with a number of free services, such as text messaging. A year later, following the dotcom crisis, budget and staff were severely downsized, and iVenus was merged with one of the company's print magazines to reduce costs.

The company chose to target iVenus on women because the company's off-line publishing titles are predominantly aimed at women. From earlier research in the US and the UK, they had expected users of their site to be young women (late teens/early twenties), often students, who would be interested in the technology but also interested in the site for the same reasons as readers of women's magazines in general – as a luxury, glossy item associated with private, pleasurable moments. In the event, contributions on the interactive parts of the site revealed that the users of iVenus were about twenty years older than expected. They read the magazine at work rather than in the evenings or weekend, and they saw it as a source of useful information and advice rather than a luxury item. They were more interested in topics such as horoscopes, beauty and fashion than technologies, yet they were very attracted to the interactive options which were subsequently expanded as a result of this feedback.

Like the Femme producers, the editors did not want to give their site a 'too feminine' look and feel. The visual designer did not want a "big girly, flowery kind of site". Colour and visuals were left as they were in the software package they had bought 'off the shelf' – as a regular, general design. Nevertheless, as with Femme, the iVenus designers chose to fill the site with mostly feminine connoted topics, such as beauty, fashion, health and horoscopes, alongside career, finance, travel and games.

As with Libelle and Femme, one of the main ways designers sought to produce a site 'for women' was by making it 'user friendly'. The predominantly men technical staff and the designer presumed that women had very limited technological skills, while the predominantly women editors presumed that women use the Internet 'better' than men – that they have a much more functional approach to technology, more interested in usefulness rather than in 'showing off' or overly fancy features; and that they have more patience in browsing and more loyally to websites. Both arguments led to the desire to make the site very easy-to-use. As one of the technical staff remarked:

> Well I guess the main thing was just ease of use cos the audience wasn't going to be the likes of myself [...] you'd have to bear in mind that it was a woman's web site, and we expected our end users not to have that much

> technical knowledge. It was going to have to be as user friendly as possible so the discussion board – if you click in – you will see that there aren't too many buttons that you can click – you either register to become a member of the boards, you can create a new topic, reply to a topic or you can enter or edit one of your replies if you said something you can withdraw or whatever.

To the surprise of all the staff, it was precisely what they considered to be the most technologically advanced option – namely the interactive discussion boards – that was used the most by the women users of iVenus.

Donna Moderna (Italy)[18]

The Donna Moderna website was founded in Italy in 1998, as the digital equivalent of the women's weekly paper magazine of the same name. A picture of the current paper magazine on the homepage of the web magazine invites readers of the website to buy the paper magazine, and advertisements earn money for the web version. This web magazine attracted by far the most visitors of the four studied, with 1,300,000 hits per month and 1,875,000 users, of whom some 90% were women and 40% also read the paper magazine. The site users are somewhat younger than the target group of the paper magazine, with ages between 14 and 45. As with the other web magazines, they have a higher than average education. About 50% have a job (and often log in from their work computer) the remainder being predominantly housewives or students.

The topics addressed in the Donna Moderna site are the same as those in the print magazine: pets, news, cars, beauty, home, cookery, family, flowers, fitness, work, fashion, psychology, health, sex, money, leisure, men, travel, books but also high-tech (meaning things such as i-pods and audiovisual equipment). Additionally, the website offers the option to discuss several of these topics on interactive boards; users also can buy books on-line, subscribe to a newsletter, send postcards, fill out user surveys, check a weather forecast or their horoscope and write an e-mail to the editorial staff. The most discussed topics on the interactive boards are news and sex; money, work and fashion are the least discussed.

The site again seems to be a mix of conventionally 'feminine' elements (e.g. the content choices) with a lay-out that is not "too flowery over-the-top pink" and simple to navigate. There are quite a few more advertisements on the Donna Moderna homepage than on the other web magazines, making it potentially more confusing, and the editors of the site

18 This section is based on Fortunati [2003c] and Fortunati [2004a].

have added many more links than the Libelle editors. Nevertheless, the Donna Moderna web magazine does give an impression of being simple to use, with lots of pictures and clear buttons.

Women's web magazines as 'assisted self-inclusion'

It would be tempting to cast these web magazines as merely an extension of paper women's magazines and, on this basis, to criticise them for reproducing stereotyped and essentialist views about women. After all, the web magazine producers we studied were all convinced that men and women have different interests, different preferences for lay-out and different navigational needs than men. The model of women's magazines, and indeed other commercial products for girls or women, have been rightly criticised for treating 'girls' or 'women' as a homogeneous group, and for concentrating on very hetero-normative themes such as relationships, family and appearance (Brunsdon 1995; Cassell and Jenkins 1998a). According to McRobbie (1991), this is in contrast to men's magazines, which divide their target group according to interest. (See, for example, the ICT magazines briefly discussed in Chapter 5.) Thus, although the producers of the web magazines described here knew that between 10% and 40% of their readers were men, they explicitly aimed their magazines at 'women', rather than 'for everybody interested in these topics'. We share the criticism of designs which reproduce gender stereotypes, and pick up this theme above all in Chapter 11. Nevertheless, there is a more positive story to be told here about the women's web magazines, in which they are cast simultaneously as rather effective instruments of digital inclusion.

Undeniably, these web magazines attracted large numbers of women. At the time of our studies, iVenus was accessed by 790 visitors per day, the Donna Moderna web magazine was estimated to have nearly two million readers, and the Libelle one had 13 million page views in a year. In all cases, the majority of users were women. In so much as a significant proportion of these women had not previously used the Internet, or were not very experienced at using the Internet, then significant numbers of women used the Internet and acquired the skills to go on doing so as a direct result of the web magazines. They were indeed more included in the Information Society.

The commercial stakes of these web magazines are evident in these figures. The producers had hoped to earn money by going on-line – directly, through advertisements on the Internet, and indirectly, with the hope that the more modern image would attract a new younger audience leading to (new) advertisers. The significance of advertising income explains why the magazine producers aimed at higher earning and higher educated women: as has been observed in the USA and Canada, "advertisers want the most bang from their bucks" (Shade 2000: 221).

So how did the women's web magazines achieve this success, both commercially and in terms of digital inclusion? In the language of inclusion needs proposed in Chapter 3, they achieved this by two means. First, they met the need for motivation to use the Internet, by providing content that was interesting and relevant to the target group. Second, they met the (perceived) need for ICT capability, by offering assistance in how to use the Internet and particularly their website. We propose the term *assisted self-inclusion* to capture the inclusion processes going on here.

The term self-inclusion was coined by our colleague Leopoldina Fortunati from her study of a non-commercial Italian website called Sensa Mascera (without mask) [Fortunati and de Luca 2003]. This website was set up in 1996 to support patients, and families of patients, with the serious illness Lupus, which can cause its (predominantly women) sufferers to look as if they wear a mask. It was part of a wider health education programme, which also included periodic talks, a course and three monthly paper reviews about the disease. Initially, the site was very basic. The founder, a specialist rheumatologist, simply loaded the paper review on the website and created the opportunity for users to ask questions of specialists on the disease. In this case, very clearly, the target group was highly motivated to read and participate in the website because the subject matter was of acute personal interest.

Another non-commercial website, Women on the Web (WOW), also had a very specific target group.[19] It was set up in 1996 by a Dutch computer scientist, as a network initially called 'the Dutch Webgrrls', with an explicit inclusion agenda: it sought to make "new media like computers and the Internet more accessible to women", to help women with these interests network with others like them. They did this by offering courses, a website and mailing lists for women about computers and related topics (see Chapter 9 for more information on this website.)

Self-inclusion is the process through which a user explores a site and learns the knowledge and skills needed to access the content. It is inclusion work performed by the user which, in the websites we studied, resulted from the relevance of the content to users, and the motivation

19 The analysis of WOW is based on van Slooten and Rommes [2003] and Rommes and van Slooten [2004].

this created to go on-line. Self-inclusion is 'assisted' when producers of the website perform additional inclusion work themselves to encourage or facilitate use. In these websites, self-inclusion was assisted by the producers – not only through the provision of relevant content, but also by supporting the access and ICT capability of their readers.

We can read the efforts of the web magazine producers as assisting inclusion by building meanings and practices that would 'speak to' their users, facilitating the domestication of the websites and development of relevant practices with respect to their use. Thus, the target women of the commercial web magazines were perceived to have 'feminine' interests, but not 'too' traditionally feminine, and this shaped design choices. The sites needed to have a lay-out which would feel appealing and comfortable to 'women', but not be too 'over-the-top pink'. Similarly, the content had to speak to the 'modern women' by including interests such as travel, money and cars alongside the more conventional hetero-normative topics. The traditional model of women's magazines targets middle class, middle aged, white heterosexual women, building on traditional stereotypes of the attractive and successful husband, the proud housewife, the caring mother and the supportive friend. The genre aiming for the 'modern woman' (like *Cosmo*) assumes that their readers have somewhat wider horizons, for a more independent lifestyle and so forth. The commercial web magazines largely built on this latter model, but stretched the 'modern woman' version of femininity to include the computer interested and competent woman. To this degree, they have helped in shifting images of both women and ICT – a theme we pick up later.

As we saw earlier, the producers of commercial web magazines, Libelle especially, also went to considerable lengths to make their websites 'easy to use' and help their readers become competent and confident Internet users. Some even assisted access by ensuring multiple platforms would work (e.g., that Mac users could get on-line without having to use Flash).[20] Arguably, the producers of the web magazines performed most of the capability and access related inclusion work for their users early on. Their aim in so doing seems to have been to trivialise the technology through which the users access the website, thus lowering the threshold for participation and 'normalising' the technology as part of everyday life. Whilst the designers and producers started from a perception of women as needing simple web designs, the increased ICT capability they facilitated will have reduced the likelihood of their women users needing

20 Incompatibility of platforms is an often underestimated aspect of software (Star 1999: 380).

simple web designs. This explains why the assisted learning we observed was probably largely a transitional effort, something that was outspoken in the case of Libelle. Yet, as the producers' efforts in this direction diminished, they were to some extent replaced by assisted learning from other users (see below).

It will be clear that producers of the non-commercial websites, Sensa Maschera and WOW, performed considerably less inclusion work in terms of capability building than did the producers of the women's web magazines. In part this was precisely because they were non-commercial: they simply didn't have the resources to commit to this. But in addition, the users of Sensa Maschera and WOW were arguably more motivated than those of the women's web magazines to do the inclusion work needed to acquire ICT capability themselves. In these cases, the relevance of the content was sufficient to recruit them, whereas in the case of the women's web magazines, the websites had to be made easy-to-use before large numbers of readers could be recruited. To generalise the point, if potential ICT users are highly motivated, they will be willing to do much of the work required to meet the other inclusion needs of access, capability and support themselves. If the motivation is weaker, then producers must do more of the work required to meet the digital inclusion needs of potential users.

Another interesting distinction to emerge from a comparison of the women's web magazines and the two non-commercial sites we studied concerns the nature of the content and relevance. Neither Sensa Maschera nor WOW started from the presumption that women are different from men, and neither sought to attract women users by stereotypically 'feminine' topics. Although we may presume that many women were attracted to the rather broad and normative content of the web magazines, we cannot say from our material just how important this content was or indeed how much women readers were influenced by it. Nevertheless, the growth and popularity of Sensa Maschera and Women On the Web demonstrate that is it not necessary to use stereotypical content in order to attract women users. Websites offering other, more specific topics can be highly motivating for particular groups of women – along the lines of the specialist men's magazines mentioned earlier.

In all cases, relevance emerges as a central element in motivating potential users: it is after all the content which attracts people to use a particular website not the fact that it is a website. This is a theme we will see time and again in this book, a more extreme case of the phenomenon described in Chapter 3: courses which start from what computing applications are relevant to trainees are far more likely to recruit and socialise the digitally excluded than courses which focus solely on computer skills. The assisted self-inclusion we identified in these websites may be said to have operated chiefly as a mechanism of recruitment: it attracted new users to the Internet and got them going. The second form of inclusion we identified emerged out of the first and operated as a mechanism of socialisation.

Inclusion through on-line interactivity

Although the women's web magazines started off as unidirectional transmitters of information, and in spite of the initial presumption that women are unconfident about ICT, interactive features proved very popular with users. In response, producers of Femme, Donna Moderna and iVenus created more and more easily accessible interactive spaces – mailing lists, discussion boards, forums and chat rooms – where users could communicate with one another, exchange information and create their own content. Significantly, users on Donna Moderna interacted on topics extending well beyond the categories prescribed by the designers. And on iVenus, a culture evolved in which users would share personal stories and experiences, and offer support in dealing with recognisable everyday life situation, such as illness, discrimination or a difficult (man) partner. Thus, the women who participated in these interactive sites were no longer passive consumers of the website, but active participants shaping the content and styles of the websites. In effect, they took over some or all of the motivating, capability building and supporting inclusion work initially performed by the producers.

We are calling this inclusion through on-line interactivity. This form of inclusion is facilitated if websites provide interactive facilities where technical matters can be discussed and where users can develop social bonds and communities over shared interests. Although social network sites such as Facebook or LinkedIn also offer these options, the goals of these sites are from the outset different than mailing lists or web forums (Donelan et al. 2009). Empirical studies of social network sites show that they are mostly used to connect to familiar people rather than with new people (Walther et al. 2008; Houghton and Joinson 2010; Donath and Boyd 2004), whereas in the mailing lists and web forums we studied users rarely knew each other beforehand and found each other on the basis of shared interests.

The two non-commercial websites we analysed were interactive from the outset. 'Senza Machera' users were encouraged to ask questions of specialists on the disease. All questions and answers were then recorded in a database, which could be consulted by other users with a search engine afterwards. Users kept asking for more interactive features as a quicker way of exchanging information about Lupus, so when a commercial firm was made responsible for maintaining the site, these functionalities were subsequently extended to include a forum and chat facility, which were used enthusiastically. The WOW strategy was to create a 'safe' space for women on the Internet and to build a sense of community amongst members – by adhering to strict rules of netiquette, by having moderators and an e-zine, and by organising 'In Real Life' (IRL) events such as courses and informal get-togethers. For a long time, WOW was the largest Internet community in the Netherlands (Drost and Jorna 2000), with some 4500 subscribers in 2003.

Some of the interactive sites addressed ICT explicitly. On several discussion boards, users discussed social aspects of the technology – such as, what age children should be given a mobile phone or how to deal with a partner's (excessive) computer use at home. And some discussion boards were expressly founded to help women acquire ICT skills – notably, the WOW boards and the computer boards in Libelle. However, the majority of the sites addressed other topics. Women participated because they were interested in the topics and because they valued interacting with others who shared these interests. So motivation was sustained due to relevance again. But in order to participate, women often needed to improve their ICT capability. Here, as in the sites expressly set up to do this, we found that women taught each other, and gained support and confidence about using computers and the Internet from each other. Where technology was not the topic, technical advice was often given casually, as an almost incidental adjunct to furthering the shared interest. On one iVenus site, for instance, several women joined in an effort to learn how to include pictures in their postings.

The presence, support and sense of humour of other women resulted in a feeling of empowerment in those women who at first had little confidence about their computing skills. As one wrote:

> Thanks girls I finally did it, I cant' believe it. I guess I'm just not as quick on the uptake as the rest of you but I'll sleep easy to-nite knowing I've now mastered it, yippee.

Similarly, a woman on Sensa Maschera wrote:

> You're right about the Forum, it took me a while to learn how to use it properly because I'm a bit thick when it comes to these things, but I'm as fast as Speedy Gonzales now!!!!

Many users of the Sensa Maschera site had never before used a forum or a chat function:

> It's the first time that I find myself writing to a forum and above all to this forum … now I am finding myself searching the Internet for anything that can help me to understand what's wrong with me.

Gradually, women became more and more expert in using the Internet, leading to what Fortunati [2004: 267] calls "a cascade process" of inclusion: "the e-mail is opened, you learn to chat, to use the forum, etc." Roughly one quarter of her survey respondents declared that, as a result of consulting Sensa Maschera, they made greater use of the Internet, and almost one fifth said it had led them to acquire greater competence in the functioning and potential of the Internet. On several boards, women voiced their pleasure in having helped another woman and later seeing how that woman becomes teacher to a new user.

Clearly, the users on these interactive sites played the role of 'local experts' (see Chapter 3) for each other – informing and enthusing them about interesting uses of ICTs, helping them learn how to use the technology effectively, and supporting them in various ways over their computer use. In all of the websites studied, women empowered each other by serving as role models. One WOW user remarked, "I have become so stubborn, I think: if others can do it, then why should I not be able to do it?" And an iVenus user said, "by reading how others deal with computer problems […] I get the sense that I could also learn it!" We observed similar processes in women-only ICT training courses (see Chapter 9). We also see in these accounts the 'multiplier effect' evident in the Ardmore Network, as the learning and confidence built by local experts begets new local experts in the community.

Many of the WOW users we interviewed found using their on-line local experts preferable to using local experts who are 'warm', in Barkadjieva's (2001) sense of being in their existing 'off-line' network of family and friends. One said, "I want to manage and maintain my own computer without being dependent on others". Relying on warm experts can create an unwanted dependency, increase feelings of incompetence and even gender hierarchies, especially when this is a man partner (Rommes 2002c). The help women gave each other on-line had all the advantages of local experts without the disadvantages of their being 'warm'. Answers were given fast and checked by many other users, who would often provide additional information. The information provided was seen as more 'trustworthy', because there were not the hidden agendas or power-relations often found in a personal relationship, and because almost all the users were women.

A defining feature of interactive sites is that users create and feel part of a community (Correll 1995; Bromseth 2006). In those we studied, the sense of community was established in various ways: by maintaining boundaries with 'outsiders', sometimes by meeting in person and, in all cases, by adhering to self-regulating informal rules regarding respectful interactions. Women users changed the discussion culture of the Internet – the 'Netiquette' – at least for the discussion boards and mailing lists they were using. Rather than pasting short answers inside another woman's mail or replying 'RTFM' ('read the fucking manual') to questions, which had been the dominant discussion style on many mailing lists before, the style on all the interactive sites we studied was one of supportiveness – 'there are no stupid questions' – one in which women took time and space to answer questions elaborately, often illustrating them with personal stories and experiences. For several women, the sense of belonging to a community on-line was so strong that they talked about being addicted to the discussion boards or of the boards being a lifeline for them. In a very real sense, the Internet had become 'their' space.

Symbolic aspects of digital inclusion

Through both assisted self-inclusion and inclusion through on-line interactivity, the websites we studied met the inclusion needs of motivation, capability, and support. But this is only part of what was going on in these digital inclusion processes. Important symbolic and identity work was also going on, transforming perceptions of women and ICTs as well as the practices involved.

By getting their women readers to go on-line, and attracting new readers to their offerings, the web magazines helped transform perceptions of the Internet from being 'a toy for the boys'. In effect, they helped build a politics of diversity around the technology (Lovell 1980) by successfully constructing the Internet as an 'object of desire' for a wider variety of users than previously, with a wider variety of interests. In the process, they also helped transform the perception of women as lacking in ICT competence. From the outset, all four commercial web magazines assumed that their target group would want to learn about ICT, but at the same time presumed these women would have some kind of 'computer reticence' which their sites designs would need to overcome.

As we saw repeatedly, the image of women and femininity as including computer competence was constructed alongside more conventional tropes of femininity. Equally, women users themselves often seemed to juxtapose in a reflexive way computer competence and stereotypical femininities. On iVenus discussion boards, women users performed their computer competence while creating a 'women's world' – a fun place "where 'girlishness' can be asserted with confidence", for instance, through the use of pseudonyms, visual signatures, narrative structures and the content of messages [MacKeogh and O'Connor 2004: 188]. Similarly, some women expressed a stereotypically 'feminine' type of ignorance and incompetence, whilst simultaneously developing their technological competence on the boards – as we saw in quotes in the previous section.[21]

Increasing confidence was a recurring theme. Writing and reading about recognisable personal stories gave many iVenus users more self-confidence, both in their lives and in dealing with technologies – again something observed in the women-only ICT courses. For one WOW user this growing confidence about technology led her to a career in ICT:

> It gives me the feeling that I belong to it. And because of this, I have started taking myself more seriously in the area of computers. Nowadays, I follow a master education media design.

What we see here is a virtuous circle linking shifts in identity and practices: increasing computer use and ICT skills leads to growing confidence which in turn facilitates greater use and capability.

Time and again our user responses conveyed a sense of belonging and of 'feeling at home' on the Internet. This resonates strongly with the whole metaphor of domestication. As the computer became normalised for these women, became part of their everyday lives, so it also became part of their self-perception and identity. The very process of inclusion work – of building enthusiasm, confidence and knowledge about using ICTs – necessarily involves changes not only in people's practices, but also in how people see themselves and the technology.

21 According to Elkjær, this way of performing femininity by girls allows them to develop their skills better than the boys in the classroom: 'the girls can develop their knowledge without anxiety the [boys] have, that of having constantly to secure their position in the public sphere because expertise in the subject content constitutes part of their identity' (Elkjær 1992).

In sum, the symbolical reconstruction of women and the Internet as 'belonging' to each other can be seen as vital part in the process of socialising women into the Information Society and of domesticating the technology. And this process will 'ripple out' beyond the immediate users. Because these websites are 'technological' and, at the same time, created 'places for women', their very existence sent the message that women can be computer competent and 'feminine'. To the degree that the websites allowed women to have fun on the Internet or to discuss deeply personal issues, they also challenged or loosened gender binaries – of women/serious/functional use versus men/fun/pleasure use, or women/private spaces versus men/public spaces. The website users' perception of themselves and women in general as being computer competent, and their perception of the Internet as an interesting place for women, was inevitably further diffused in wider society, through their on- and off-line communities.

Conclusions

We have shown that websites can be important instruments for digital inclusion. Those described here all offered one of the first places 'for women' on the Internet in their respective countries. In numerical terms, they increased the inclusion of women in the Information Society dramatically. Hundreds of thousands of women started using the Internet, and/or used it more intensively and more competently, as a direct result of these websites targeting women. Where previously the Internet was just one more 'toy for the boys', their early web presence contributed to making the Internet what it is now – a place where diverse users can find information that interests them and meet others who share those interests. We can draw out a number of generalisable themes from the analysis in this chapter.

First and foremost, in studying how these digital inclusion outcomes were achieved, we have identified two distinct types of inclusion processes: assisted self-inclusion and inclusion through on-line interactivity. Self-inclusion implies that the user is motivated to do a fair amount of the inclusion work: they are attracted to the website because the producers have succeeded in making it an 'object of desire'. In the case of the commercial web magazines, this seems to have been due (initially) to the appeal of the rather general content. In the case of the two non-commercial websites, users were attracted because of a very specific interest – they had a serious illness or they felt themselves to be in a minority as women working with or interested in computers. In all cases, this self-inclusion by the users is nevertheless assisted to varying degrees by the producers. They performed inclusion work not only through their provision of relevant content, but also through the various forms of guidance in how to use the Internet and computers. The latter involved considerable effort on the part of the web magazine producers; obviously motivated to facilitate the access of potential readers to their sites. By contrast, the two non-commercial sites relied more on users to do the inclusion work needed to learn how to use the Internet – who, in any case, were probably more motivated to do so than were users of the web magazines.

Where assisted self-inclusion operates as a mechanism of recruiting people into the Information Society, inclusion through on-line interactivity may be seen as a means of socialising people into the Information Society – of sustaining and continuing to support their engagement with ICTs. It occurs through on-line discussion boards, mailing lists, forums and chat rooms. These interactive sites revealed many features found in other inclusion strategies we studied: the role of local experts in facilitating the build up of ICT capability; the importance of mutual support and encouragement; how empowering it can be to become ICT competent; and the mutual reinforcement between this and increasing self-confidence in one's life. In effect, by supporting learning and generating interesting content, users take over much of the work necessary to meet the continuing inclusion needs of motivation, capability and support.

The central importance of relevance cannot be overstated, in both forms of inclusion. Self-inclusion occurs because users are attracted to and interested in the topic(s) of the websites – because it has some relevance in their lives or some appeal in terms of their identity. In the same way, inclusion through interactivity highlights the dual importance of both content and community: it is the shared interest in a topic which motivates users to participate in an interactive site. As we argue in Chapter 9, there is often a 'double articulation' going on here, where content enhances community and vice versa, in a way that is highly productive in terms of digital inclusion.

Investigating who does what inclusion work has been instructive. Comparison of the commercial and non-commercial websites revealed that the balance between the inclusion work done by users and by producers may vary. In this case, users of the non-commercial websites were so motivated to use the site that they were more willing to do the work needed to acquire ICT capability themselves. This indicates a significant merit of our focus on who does what inclusion work: it reveals important interconnections between how the different inclusion needs are met.

A second insight from this focus on inclusion work has been to unpack the process of domestication. As we saw with the web magazines, the more inclusion work is performed by producers of a site – in developing content that fits with the desires and needs of potential users, and in designing the site in a way that fits with users' skills, preferred way of learning and the hardware to which they have access – the more the technology itself is trivialised or made invisible to the user. This strategy has obvious benefits for digital inclusion, in so much as lowering the threshold of effort and knowledge required to go on-line enables more people to use the technology. It also means that the uses and applications – what generally attracts and motivates most people to use ICTs – are foregrounded.

Nevertheless, there are potential downsides to an inclusion strategy which trivialises ICTs. First, the less inclusion work users do, the less knowledge and skills they may acquire about the technology, which may limit their ability to participate more fully in other aspects of the Information Society. Second, making the technology invisible also renders invisible what choices the designers have made for the users and 'the paths not taken' – how users could use the technology differently. As Crutzen puts it, the critical or 'transformative room' is small in these cases (Crutzen 2007). Similarly, one of the main points of feminist and queer critique of social network sites such as Facebook and LinkedIn has been that users are offered fixed categories and identities to select from. By offering some categories and choices (e.g. man/woman; heterosexual/homosexual) and not others, these identities and certain kinds of subject formation become normalised, and it becomes less visible that these categories are merely designer-made constructions (House 2011; Light 2011).

However, in generating new content and asserting new discussion styles, women on the interactive sites we studied were appropriating and utilising the technology differently. This is precisely what we anticipated in our introductory chapter: we posited the Information Society as a 'work in progress', and suggested that increasing the diversity of those using ICTs would in turn have an impact on the shape and direction of that work in progress.

Finally, the impact of the websites described here can be measured not only in the practices of the women users – their increased ICT use and competence – but also in their perceptions of themselves as computer competent and of the Internet as interesting. Arguably, the wider symbolic impact of these identify shifts is considerable, transforming perceptions of both femininity and the Internet as things which can 'belong' together. This observation serves to remind us that symbols, identities and practices are intimately interwoven and (to some degree at least) mutually reinforcing in social life. Thus, the early increase in women using the Internet helped destabilise the perception of the Internet as 'toys for the boys', and this changing image in turn contributed to women and men now going on-line in more equal numbers. We see here the importance of changing both gender(s) and technologies, – as anticipated by our 'co-production' framework. We also see the importance of both increasing numbers and symbolic transformations as critical elements in digital inclusion efforts, where both kinds of changes (numbers and symbols) are simultaneously an outcome and part of the process of including more women into the Information Society. This is a key conclusion of this chapter, to which we also return in Chapter 10.

5

Fun and play in digital inclusion

Helen Jøsok Gansmo

The previous chapter analysed how web magazines for women provided opportunities for 'self learning' and 'self inclusion' by providing interesting, relevant and easy-to-use web pages. 'Relevance' refers to a variety of content, a mixed bag of information and entertainment. A long-standing belief sees women as particularly concerned with the useful aspects of ICT. This belief was rendered problematic in the previous chapter, where we saw how fun and glam were important ingredients in web magazines for women. In the framing of inclusion needs introduced in Chapter 3, this chapter examines and reflects critically on the potential role of such ingredients in providing the motivation to get involved in using and/or designing ICTs. It asks: Are fun and play effective routes for including girls and women into the Information Society? Do promises of enjoyment make ICT more attractive to girls and women?

Fun is no straightforward issue, as is elegantly captured in the words of the Danish poet and designer Piet Hein on The Eternal Twins:

> Taking fun as simply fun and earnestness in earnest
> shows how thoroughly thou none of the two discernest.

Actually, fun and earnestness may be two sides of the same coin. As we show below, much of the research on gender and ICT has painted a dichotomous either-or picture – the boy or man computer user is portrayed as having fun while the girl or woman user is portrayed as only interested in the potential utility of the computer. This literature is used as a backdrop for the body of the chapter, which analyses several different efforts to make ICTs more attractive and enjoyable for girls or women by encouraging them to have fun with ICT.

Previous research on gender, ICT and fun

Research on gender and ICT appears contradictory on the subject of fun and inclusion. On the one hand, there has been a widespread perception that, compared to men, women approach technology with more critical questions and a stronger emphasis on issues of usefulness and relevance (e.g., Whyte 1986). By implication, fun would seem the wrong way to go for the purpose of including more women into the Information Society. On the other hand, studies of computer hackers demonstrate that enthusiasm and fun has been central for developing a strong interest in computers in this group (Turkle 1984; Nordli 2003). Sherry Turkle's work has been very influential in bringing a gender analysis to this subject. Drawing on Weizenbaum (1976), she painted a picture of the hacker computer users as young men who have an intimate, passionate, all-encompassing relationship with the computer. This image has informed a perceived link between enthusiasm and competence. Gansmo et al. (2003a) use 'boys' room competence' to label the competence which lonely and isolated nerds are believed to acquire from enthusiastic and absorbing tinkering on their computers. Turkle (1984) stressed that the a-social hackers seemed to 'love the machine for itself' but often to the exclusion of loving humans. Later, Turkle (1988) argued that women would be put off by this model of computing and want to stay away from the 'intimate machine' as a result; arguably, the computer represented a personal and cultural symbol of what a woman is not.

It is an open question as to whether this remains the case, given the profound changes in available technologies and practices in the last 20 years. However, as Tine Kleif and Wendy Faulkner (2003) argue, it is significant that being technology-focused is often perceived as mutually exclusive with being people-focused. This mutually exclusive character of the technical/social dualism maps readily onto presumptions that women are 'more social' than men, which underpins Turkle's assertion that women will tend to reject the hacker model of relating to computers.

The technical/social dualism may also contribute to making technology and engineering more generally feel like a 'gender inauthentic' career choice for women (Faulkner 2000a, see also Chapter 2). But does it also make having fun with ICT as 'gender inauthentic' for women?

In some ways Kleif and Faulkner (2003) seem to think so, though they do not address ICT specifically. They point out that enthusiasm and pleasure with respect to technology is a topic that largely has been left unexplored in both feminist and technology studies. Kleif and Faulkner further ask why it is that some men take such pleasure in technology while many women either do not experience such pleasure or tend not to own up to them if they do. One reason undoubtedly concerns how children's play is gendered. On the average, boys and girls are socialised to play with different types of toys leaving boys more likely to acquire and enjoy skills in hands-on tinkering with technological artefacts. Another more subtle aspect may be that boys grow up with more 'permission to play', while girls are more likely to be socialised to devote more thought and time to care for others. Switching off from responsibilities by having fun could thus be seen as a more available or legitimate option for boys and men than for girls and women. Women may perceive fun with technology as a gender inauthentic pursuit, and they may need to legitimise or justify their leisure activities more than men do (Kleif and Faulkner 2003).

There is some evidence that supports this claim with respect to ICT. Haddon (1990) found that girls who took pleasure in the use of computers could not relate to it just as a toy like boys could: they usually had to justify their interest in computing in terms additional to any pleasure the activity might provide. Similarly, Turkle (1988) argued that women totally reject such pleasure. However, these studies may now be dated. Several more recent papers suggest that with respect to computer game playing, women are as motivated by fun and pleasure as men. What differs is what women and men consider being fun and pleasurable, although this may depend on the context of play (Carr 2005, Hartmann and Klimmt 2006, Jansz et al. 2010, Williams et al. 2009).

Anyway, a gender dualised toy/tool dichotomy has been widely assumed – that boys and men regard computers as toys to play with, while women and girls use them as tools to do things with. It has been argued that the boy/toy vs. girl/tool assumption has been reinforced by many cultural and educational cues, all of which reinforce computing as a masculine domain symbolically (e.g. Margolis et al. 2000). Accordingly, research in the 1990s was rather unanimous in describing, and more importantly, in supporting women's instrumental and pragmatic approach to and application of ICT (e.g. Rasmussen 1997; Whyte 1986). Girls tend to be confident although pragmatic computer users when studies or jobs require the use of ICT (Durndell et al. 1995). Girls also want to use ICT to make learning more interesting. They are motivated by observable usefulness (Håpnes and Rasmussen 1997), where their computing is seen as useful to society (Margolis and Fisher 2002). Consequently, it has been widely assumed that girls had to be convinced that it was useful rather than fun to use a computer. This belief was, for example, evident in the Women and Computing Initiative described in Chapter 7.

Although much of the literature on gender and ICT has found enthusiasm to be a basis for men's engagement with ICT, this research has often adopted a rather judgemental standpoint – assuming implicitly that women's (presumed to be) more instrumental and utility-oriented, even duty-based, relationship with the technology was healthier and more sensible than men's (see Chapter 2). The (presumed) hacker or nerd model of relating to computers was pathologised, and pleasure and enthusiasm about computers was constructed as a problem which women should avoid! To some extent, this standpoint encouraged 'acts of purification' to exorcise 'desire' from the computer.

At the same time, there has remained a strong belief that boys' fun engagement with computing contributes to men's dominance in ICT occupations. And indeed, several studies show that girls have been less interested in playing computer games than boys have (Buckingham 2006; Vaage 2002). Although the relation between boys' computer gaming and men's dominance in ICT is hard to test, it has been argued that if there were more games for girls, they may gain more confidence in their computing skills (e.g. Livingstone and Bovill 1999). And there have been calls for the ICT sector to develop products for girls in order to attract more of them to use computers (Newman et al. 1995) and to entice them to regard computer science as an attractive career choice (Gorriz and Medina 2000).

Simultaneously with the focus on the apparent relation between boys' ICT enthusiasm and men's dominance in the ICT industry, the entertainment software industry has grown rapidly, with the game market for girls seen by some to have considerable business potential (Kafai et al. 2008). Since the early 1990s several companies have attempted to market a line of software games specifically for girls. Many of these companies provide girls with fun and meaningful computer experiences intended to encourage them to prepare for the demands of a technologically advanced future (Gorriz and Medina 2000, Jenkins and Cassel 1998). Some attempts to devise special 'girl software' intended to tap into girls' interests were moderately successful while others resulted in software that was less interesting or that reinforced 'feminine' stereotypes (Kafai et al. 2008). This is discussed at length in the next chapter.

Volman and van Eck (2001) argue that girls prefer to see the usefulness of what they are supposed to learn. Rather than making software especially for girls, they suggest that applying the technology in ways that girls find authentic and realistic, can contribute to gender inclusiveness. Moreover, many such interventions seem to attempt to recruit girls by offering them a 'feminine' technology based on assumptions that girls constitute a homogeneous group distinct from a homogeneous group of all boys. Thus, efforts to grant girls and women the same rights as boys and men (access to technology), also presume that girls/women's values and experiences are different from boys/men's – they have the right to a different, preferably useful and relevant, technology (Gansmo 2006).

For all the emphasis on rights and inclusion, such interventions are largely counter-productive in that they reinforce rather than challenge gender dualisms (as we show in Chapters 4, 6 and 9). Moreover, many efforts to include girls and women into the Information Society seem to have ignored play, enthusiasm and fun as possible routes to digital inclusion. They are characterised by a dualised way of thinking whereby women mainly are associated with utility and a serious use of the computer, while men are conceived as playful and tinkering. Throughout this book, we have been critical of applying binary gender categories when analysing technologies, on the grounds that such dichotomous accounts oversimplify and misrepresent both gender and technology and tend to reinforce gender hierarchies around technology (see also Faulkner 2000b, Sørensen and Nordli 2005). The remainder of this chapter investigates what happens when inclusion strategies eschew these assumptions and revolve instead around fun and play. As well as considering whether this is an effective means of including girls and women into the Information Society, we ask: can fun and play work to break out from the gender dualised understanding of fun with computers? These questions are investigated by analysing several different inclusion efforts, with a special focus on co-constructions of fun ICT and gender.

One place to have fun and another to do 'proper' computing

For more than 20 years it has been an explicit political goal in Norway to implement ICT in schools in ways that reduce the digital gender divide. How does this strategy relate to girls having fun with ICT? This question was explored through interviews with policy makers, educationalists and school pupils in Norway [Gansmo 2003a, 2003b, 2004]. According to the interviewed policymakers' accounts, boys are addicted to computer games and are thus becoming technically skilled through fun, while the previously absent girls increasingly enter the arena as the Internet has developed and activities such as on-line chatting have become common. Although girls master 'normal' computing and are seen as having some sort of fun while using the Internet, this kind of fun is not believed to trigger technical enthusiasm and interests. So here again we find dichotomous and hierarchical representations of 'feminine' computing in contrast to 'masculine' computing.

Norwegian schools and municipalities have the prerogative to decide how to carry out ICT policies. Thus, individual secondary schools were found to respond differently to the focus on the digital gender divide. To varying degrees all the schools investigated tried to include all pupils as 'normal' computer users – mastering basic application – but none applied a particular gender inclusion strategy and none applied a strategy focusing on fun. We learnt from interviews with pupils aged 14 to 16 that this made them draw a clear distinction between 'boring' school computing and 'fun' leisure computing. This 14 year old girl was typical:

> Computing is boring. It is nice to have when you have a job or if you know how to do it. But I don't find it interesting. There is nothing I would like to try. I found that out after using computers in school. We learnt where all the keys on the keyboard are. Afterwards I went home, on the Internet and spoke to other people there. The only thing I need computers for is the Internet and e-mail, chat groups and stuff like that. We're not allowed to chat in school. In school we have projects and assignments where we are supposed to learn the keyboard.

The teenagers mainly associated school computing with typing, to gather useful information from the Internet and make presentations. Thus, in their 'official' accounts, ICT was linked to useful tasks which they saw as boring because the tasks allowed no play and fun. By contrast, out-of-school computing was mainly constructed as fun, only to a small extent as useful. Leisure computing includes game playing, chatting or e-mail communication, surfing the Internet for information about 'useful issues' such as games and pop groups, and downloading music from the Internet.

We were fascinated to discover that downloading music was not presented as 'computing' by these teenagers. Listening to music was just something they did in addition to other tasks, something that went on in the background, making the technology translucent in a way that made the activity visible (or, rather, audible) while the technology became invisible. Downloading music seemed to be a major preoccupation for both girls and boys (see also Heimsvik et al. 2005). The significance of this has to date been underestimated in discussions about ICT use and implementation, but music may be changing the meaning of ICTs for young girls, and so be working as a de facto inclusion mechanism. In a similar way, the use of chat and social network sites may become fun, gender neutral activities – in which the technology becomes invisible or trivialised, in the sense that it is taken for granted and treated like a black box never opened. These activities might serve as an entrance to wider computer use and increased ICT skills.

The teenagers we interviewed in Norway did to some extent see the importance of school computing as 'proper computing', as something different from leisure computing. But they saw school computing as only one of many possible applications, because they had domesticated the 'fun' computer at home and brought their experiences from this computer use to school. In earlier studies (e.g. Håpnes and Rasmussen 1997, Gansmo 1998, Nordli 1998) teens regarded the computer as intimately linked to gender both symbolically and in terms of numbers – more boys than girls were seen to use computers, hence computing was regarded as a 'masculine' activity. Nordli [2004] and Gansmo [2004] show that computing per se is no longer intimately linked to boys, even if certain aspects of ICT are associated more with boys and others more with girls. In particular, the teenagers in the schools' study did not co-construct school computing and computing skills with gender at all.

It would seem, then, that the Norwegian policy makers and educationalists have succeeded in making computing in school transcend gender. Ironically, they have achieved this by making computing being experienced as so boring that none of the pupils really wanted to be identified with school use of computers. Schools seemed to have exorcised 'desire' from the computer. The enthusiastic inclusion to a much wider computer use takes place in the pupils' leisure spheres. Even though leisure computing still carries some traces of dichotomous gender connotations –

presumptions of somewhat more chatting by girls and somewhat more gaming by boys – these connotations do not seem to be excluding of either girls or boys. Rather, some girls and boys were seen as interested in gaming, some girls and boys are regarded as more interested in communication, and the activities were not considered mutually exclusive. Hence, the teenagers co-constructed multi-vocal representations of gender with multiple computing possibilities. Thus, the interviewed teenagers were included into the Information Society more through their own appropriations and redefinitions of the technology than through any policy strategies and school-level implementation. And fun seemed to play a crucial role.

Still, schools may have played an important role in demystifying the computer and instilling basic ('normal') computer skills. Through their inclusion of all pupils as computer users, some of the strong gender-based constructions of computer users probably have faded away. In this regard, the Norwegian schools' gender inclusion policies may be seen as a success. On the other hand, the school strategy did not seem to inspire more students to study higher level computing – neither boys nor girls. The teenagers saw school computing as related to future education in or jobs within ICT, which they mainly regarded as boring.

To a large extent, then, the findings about the teenagers' leisure computing challenge dichotomous understandings of gender in relation to ICT use. The question remains whether the diversification of leisure computing into fun activities may be regarded as means to generate more enthusiasm for 'serious' ICT? Or does the transparency of the fun leisure technology – where the technology is made invisible behind the activity it is used for – mostly work to include more people as computer users rather than as ICT students? The teens were not having fun while tinkering with the technology as in Turkle's (1984) analysis – they were having fun while applying the technology for everyday activities such as listening to music and communicating with friends.

Fun **and** *social computing as a way to break dichotomies?*

As noted above, early studies of hackers found that enthusiasm and fun were important preconditions to develop a strong interest in computing (Turkle 1984). The studies also showed that most women were put off by the observed a-social hacker model of relating to computers (Turkle 1988). Hence, a gender dichotomy between men and women seemed to map readily onto being technology-focused and people-focused, respectively. In contrast, Håpnes and Sørensen (1995) showed how a hacker culture in practice could be quite social – involving strong social interaction around the technology and not just through it. In the hacker community studied, the young men took care of each other as well as socialising, and Håpnes and Sørensen argue that this was compatible with presumptions of a 'feminine' social culture.

The social character of computer enthusiasm has been observed on a grand scale at large computer parties of which The Gathering in Norway is Europe's largest. We did a study of the The Gathering, interviewing organisers and participants (women and men) in 1999, 2000 and 2003 [Nordli 2003a, 2003b]. At these computer parties, more than 5000 young computer interested people, mainly 15 to 20 years old, got together around their computers for five days and nights at the same location. The organisers' aim was to make the most of the social aspects of computing by creating a space in which young people who enjoy computers were encouraged to meet and interact. In the process, they meant to reconfigure the image of computer users. And arguably, the most important inclusion impact of The Gathering is that it conveys computing as a social and fun activity, and so contributes to breaking down the stereotypes about computer enthusiasts.

This is an inclusion strategy directed at young women as well as men, and the attendance of young women has grown to around 15-20% in later years. The inclusion efforts in The Gathering had a dual focus. They tried to include competent girls in the crews responsible for organising the huge event. This inclusion effort only reached a few established computer enthusiasts. The wider inclusion idea was to attract as many different types of young people as possible to participate in the party. This was achieved by various means: tickets were reasonably cheap, the party was safe (no drugs/alcohol allowed), many different activities were available, and the party was promoted as cool, fun and social. Recruitment and socialisation into computer enthusiasm was supposed to work through the experience of being a participant; recruitment to the party depended in large measure on what participants said about it on their social grapevines. As one 17 year old girl told us:

> I've heard a lot of rumours about how much fun people have here. Quite a few of my girlfriends have been here before. This year I thought; finally it's my turn to join the party!

The goal of The Gathering to reconfigure the image of computer users by encouraging social interaction was obviously appreciated by the participants. Nordli [2003a, 2003b] observed that they came to talk, hang around, get new friends, to dance and have fun. The main computer activities were computer games, chatting and downloading, especially of music files. Usually they spent 3 to 4 hours in front of the computer each day during the party – the young women and men alike. The rest of the time they spent talking to people or participating in other activities. Interestingly, they reported that, even though their common enthusiasm for computers was what brought them together, they usually sat in front of the computer only when they were tired or bored.

When asked about gender differences all the interviewed young women and men stated that they saw no real reasons why computing should attract more men than women. Yet strikingly, their accounts of their own and others' computer use were clearly gender differentiated. Girls were seen to be interested in chat, while boys tended to be seen as more interested in configuring the computers and programming. Significantly, their actual practices were far less gender differentiated than their accounts. Very often the young women and men at The Gathering were doing the same things on their computer, but they described their activities differently according to gender dualised representations of 'masculine' and 'feminine' computing [Nordli 2003a]. To some degree then, the gender difference seemed to be more about how they accounted for their computer activities than what they actually did.

The gender dualised representations were associated with assumed gender differences in skill levels as well as preferred activities. In general, boys were presented as being more skilled than girls. This is in part because of the low status given to the activities that girls were assumed to engage with. In spite of The Gathering's emphasis on socialising and having fun, the participants did not consider chatting or designing websites as skilled activities. Computer skills were rather associated with programming and the configuring of computers. Interestingly, none of the girls or any of the boys we interviewed had any knowledge of such skilled activities! The participants at The Gathering seemed to follow an established gendered hierarchy of attribution, which gave more credit to the young men's fun than to the young women's. Paradoxically, there seemed to be a tendency to regard what the boys were doing as 'serious fun', while what the girls did was 'just fun'. Game playing, a more typical activity for boys, was given a lot more credit and status than girls' fun and play with chat and web design.

Although The Gathering succeeded to some extent in creating a space for girls as well as boys to have fun and to socialise around their computers – thus challenging the binary association of girls with useful computing only – these gaming communities apparently failed to transgress the asymmetrical character of gender. The young women found it difficult to be taken seriously as computer enthusiasts. Most of the boys and young men interviewed took it for granted that the girls and young women were only interested in communication or chatting up boys. Thus, to be a girl or young woman and seriously into fun computing seemed difficult because of such gendered expectations. These expectations may be conserved through the continued references to the myth of the hacker. This is why Gansmo et al. (2003b) propose that feminist studies of ICT should stop using the hacker concept because too often, it is used as a stereotype that reproduces rather than changes gendered perceptions of computing (see also Chapter 2).

The organisers of The Gathering displayed a kind of evangelising attitude in relation to computer enthusiasm: they wanted to share with as many people as possible the message that 'Computers are wonderful'. Arguably they have contributed somewhat to changing the image of the nerd and hacker, and to making computer enthusiasm more attractive to more women. But still they struggle to attract enough young women and have them recognised as participants on an equal footing with the young men because of enduring gender dichotomies about having fun with ICT. In addition, there remained quite strong views on the gendered distribution of skills in the culture of the computer party. So, whilst The Gathering had some success in challenging the girl/tool vs. boy/toy dichotomy, this was replaced with a new gender dichotomy whereby the new 'feminine' technology use was seen as trivial or 'just fun' in contrast to a 'masculine' technology use of 'serious' fun which continued to be seen as more skilled.

It is worth reiterating that in spite the enduring recourse to hierarchical gender dichotomies that we see time and again in the interviewees' accounts and representations of what they do, the actual practices of participants in The Gathering were generally diverse or not gender differentiated. Thus, the emphasis of The Gathering on encouraging social interaction was obviously appreciated by the interviewed participants, all of whom went there to socialise. Yet, this social feature seemed to attach to the girls participating, not to the boys. And even though the participants at The Gathering played with their computers in many different ways during the party – as they do in their leisure time elsewhere – this community of computer enthusiasts confirmed that the discourse related to the presumed importance of 'boys' room competence' still was effective (see also Gansmo et al. 2003a). While many girls were included as users, they were not acknowledged as 'serious' users. Transgressing this appreciation of rather particular skills and enthusiasm remains a challenge for the inclusion of girls and women into the Information Society. Fun and social aspects of computing did seem to break some dichotomies, but new dichotomies emerged which continued to value 'the masculine' over 'the feminine'. In a way, discourse lagged practice.

A further point bears mention. Vivian A. Lagesen [2004] notes that the young women ICT enthusiasts at The Gathering were rather pragmatic in their enthusiasm. Contrary to Haddon's (1990) claim, they perceived the computer as a tool to do fun things with. But there was nothing special about the computer to these young women. Much as we saw with the teenagers interviewed in the school study, the computer was part of their everyday life, a trivialised artefact they used to pursue their favourite activities of chatting, making web pages, downloading music or playing games. None of the young women interviewed at The Gathering planned to study computer science. This was not because they feared computers or expressed any anxiety that computer science was a 'masculine' domain. Rather, some actually wanted to work with something else in order to be able to maintain their pleasure in leisure computing. They were afraid that if they turned their main interest into a job they would become fed up with it. Others claimed not to find computing that interesting; or they found other career opportunities more tempting. So, even among the more computer enthusiastic women, the technology was in a sense getting trivialised: as computers became a natural element in their everyday life the technology became taken for granted. In a way, it was almost invisible while the activity it was used for was foregrounded.

Although fun can be very important to the inclusion of girls and women as ICT users, such fun inclusion does not automatically lead to an interest in ICT as a career. The question then is whether fun may be used explicitly as a strategy for including young women into ICT occupations.

Fun as a gateway to education?

Several studies have shown that a career in ICT has very negative connotations. Someone working with ICT is represented as a 'nerd', 'geek', or someone with a boring 'trainspotter' mentality. ICT specialists are viewed much like the hacker enthusiast: whilst they are intelligent and have great interest in computers, they spend most of their spare time in front of monitors and (so) are short on social skills (Kvaløy 1999, Gansmo 1998 and Nordli 1998). To the extent that this image does hold and that it does deter young women from a career in ICT, it is clearly a major barrier to be overcome. A UK government-backed initiative called IT Beat sought to drastically change the image girls have of geeky men in ICT by creating a space where girls could realise the fun and glamour of computing and so making it relevant to their lifestyles and aspirations. Lisa Pitt studied this inclusion strategy through interviews with the organisers and participants of the initiative and through observation at the slumber party [Pitt 2003a, 2004a].

IT Beat initially encouraged girls aged 11 to 15 to acquire web design skills through a competition where they were guided through a process of designing a fan website for their favourite pop band. The IT Beat website and the competition offered large amounts of information, support and advice on how to create the fan sites. Girls of varying ability and knowledge could apply this information in their own time and manner. Through the competition the girls were turned into active learners. 250 web design contributions were rewarded with tickets to what was heralded as the world's biggest slumber party. This was held in the Science Museum in London in 2003 with a pop concert as the main event, along with competitions and opportunities to meet a lot of glamorous women ICT professionals as role models and chaperones.

The strategy was to break out from the girl/tool vs. boy/toy dichotomy by focusing on girls AND fun, and to attract the girls by starting from established girls' interests in popular music and pop stars. Instead of focusing on the technology, IT Beat tried to entice ICT inexperienced girls to have fun through the indirect use of ICT for something that was of interest to them. The key idea was to reconcile learning with entertainment. Buckingham and McFarlane (2001) assert that this approach to learning and inclusion is rarely used and is often treated contemptuously, yet it has been found to be a good way to re-engage individuals and make technology relevant to their lives. Pitt [2003a] asserts that the same applied in the case of IT Beat: fun and relevance were critical to the success of the strategy. As one of the organisers of IT Beat argued:

> If you want to make something happen, you have to appeal to what they're actually interested in. It's no good thinking about what they should be interested in... As opposed to the old route – teach kids because they have to learn, I want kids to learn because they desperately want to, because they are doing it in a fun way ... If you're trying to make something more attractive, then you got to make it seem more fun. And that's exactly what we're doing

> ... You can't change the image of an industry by saying actually it's all very boring. Because that's what they already think, so you've got to take it to the other extreme.

The slumber party itself, and indeed the competition, made leisure and entertainment technological, and thus created a fun image of technology. It seemed a seamless integration of web talent and pop culture; because the content was fun, the technology could become fun too [Pitt 2004a]. This incorporation of technology into leisure through the pop-oriented content echoes the inclusive impact of downloading music for teenagers. Arguably, this is precisely the kind of design strategy likely to encourage girls and women to use ICTs. Indeed, work on leisure, gender and design has found that "women might make more use of computer-based technology for leisure if it is designed to fit into existing routines of leisure activity" (Martinson et al. 2002:29).

The success of the IT Beat strategy in terms of career choices will not be known for some time. Nevertheless, the fact that they managed to captivate the imagination of 250 girls and to keep them enthusiastic throughout the night shows how accurately they understood the interests and aspirations of their target audience. Having said this, there was a clear preference for pop stars over the ICT activities throughout the slumber party. One of the difficulties with an initiative such as IT Beat is to determine how much the girls will be able or willing to link the fun and glamour of the whole event to a career in ICT. The poor representation of women in ICT is a long term problem that cannot be solved with a quick fix. Yet, the subtlety of the approach may in fact be better suited to the delicate situation and may indeed prove more effective than a strong handed one [Pitt 2003a, 2004a]. Enabling girls to enjoy computer technology now may be more effective than explicitly encouraging them to think of ways in which they can apply ICT knowledge professionally in the long run.

But in the IT Beat strategy, as in leisure computer use more generally, we see a process in which the technology is made trivial as the activity it is used for is foregrounded instead. Thus, in the very design of the inclusive activities of IT Beat, the technology was deliberately kept in the shadow of the glamour of the pop stars. This is a familiar phenomenon in domestication theory: technologies that are widely and obviously used in everyday situations (such as television and telephone) are hardly perceived as technology by its users (Silverstone 1994, Lie and Sørensen 1996). When computers are made invisible in this way, they may be severed from their masculine, nerdy image. And certainly, we observed that when the girls found the technology enjoyable or fun, they no longer thought of it as technology. For instance, one young girl who mentioned how much she had enjoyed trying to mix music on the DJ turntables was asked if she thought of that sort of activity as related to technology; she simply said “Oh no, that’s not technology – that’s just fun.” By implication here, these girls still associated ‘technology’ with something alien to their lives. Yet it is clear that the girls did enjoy using technologies and (in many cases) had gained greater skill in using them, even if they no longer at that point saw them as technologies.

In the case of both IT Beat and The Gathering, the image of computing as fun for boys but not for girls was clearly challenged. The IT Beat strategy confronted this image by creating a space in which teenage girls could have fun with computers in the context of their interest in ‘boy bands’. In this way it also conveyed that you can be a girl in stereotypically feminine ways (liking boy bands) as well as in stereotypically masculine ways (having fun with computers). The two are not mutually exclusive. Attempts such as the IT Beat may thus serve as an example of potentially progressive gender changes for the girls having fun with ICT. Certainly IT Beat appears to have hit their target audience spot on by making ICT relevant to teenage girls’ pop mania, and by showing many girls that ICT can be fun and glamorous. But the strategy may run the risk of not nurturing the desire for learning more about the technology itself to the same extent.

Women just want to have fun?

The studies discussed in the previous section were primarily concerned with girls and young women, showing how leisure computing may be considered fun and how more playful inclusion strategies may be effective. Moreover, as we saw in the previous chapter on web magazines, also adult women may be motivated to use computers because they provide access to entertainment and leisurely activities such as chatting. Does this mean that the situation now has been turned up side down – so that girls and women only want to have fun with computers?

We may begin to address this issue by comparing the web magazine Femme, presented in the previous chapter, with HjemmePC and PcPro – two web magazines produced for readers interested in computer-related information, a group traditionally believed to be men [Hestflått 2003]. PcPro focused on the newest applications and entertainment for more advanced computer users. HjemmePC was directed at the ordinary home user with its more educational content oriented towards utility.

HjemmePC tried to reach out to all computer-interested, non-specialist people and claimed to have many women readers.

As we saw in Chapter 4, Femme was defined as an extension of a traditional printed woman's magazine with traditional entertainment columns such as fashion, beauty, family issues and food-recipes, celebrities and career. Also PcPro stated that one of their goals was to entertain the reader. Arguably, they used an extended concept of entertainment, given the kind of content they offered: the PcPro reader was supposed to be entertained by information about new technology, gadgets and technological accessories. This was somewhat different for the reader of HjemmePC, who was expected to be entertained through education, possibly also by reading about entertainment such as new computer games. Thus, entertainment was central to all the magazines. However, while the entertained readers of HjemmePC and PcPro were provided with knowledge they might use to get even better included in the Information Society, Femme's readers were apparently just entertained.

The editor of Femme opposed the strong conception of women as only interested in useful aspects of ICT. She saw women as wanting to enjoy the technology rather than simply using it instrumentally. Hence, Femme added interactive services. Unlike HjemmePC and PcPro, Femme emphasised the importance of 'community building' and thus offered MUDs that allowed their users to generate their own content. As we showed in the previous chapter, this provided women with opportunities of self-inclusion that the computer magazines did not offer. Also the readers of the Irish women's magazine iVenus, described in the previous chapter, told about their participation in the magazine's on-line forums that joining a community where feminine 'girlishness' was validated, provided a lot of enjoyment.

However, the women iVenus readers also described experiences similar to the women at The Gathering. Using the discussion board was a stigmatised activity devalued by outsiders. Some women took on board this denigration of their activity by referring to their 'addiction' to the forums much like the ways in which women have referred to their relationship to other 'feminine genres' like television soap-opera [MacKeogh and O'Connor 2003]. This indicates that participation in such women's discussion forums could be doubly stigmatised: not only did the interviewed women feel that there was a stigma attached to the notion of having on-line friends, but also that they were seen to use the on-line forum inappropriately as a 'girlie site'. However, the dismissiveness of their men friends did not appear to keep the women from using the forums. Actually, this dismissiveness may have been pretence. We know that the similar forums offered by Femme ended up also attracting a lot of men users. There is little doubt that both men and women are interested in communicative fun through the use of ICTs [Spilker 2003].

Thus, the social and community aspects of ICT seem to have several effects. First, to be social around the computer may serve to reduce the negative image of the 'lonely nerd'. Second, 'to be social' was also somewhat negatively associated with a nattering women's culture. But the forum users we have interviewed show that, no matter the construction of fun ICT as either means or end, they were able to transgress such distinctions and have fun both with and through ICT. Third, to be social with, through or around ICT also offers ample possibilities for learning both 'on demand' and as a consequence of the initial social activity, as we saw in the previous chapter. Hence, the women users may actually learn a lot while having fun and may be motivated by this learning. Thus, it would be misleading to conclude that women only want to have fun with computers.

Toy and *tool, fun* and *useful, boys* and *girls*

The inclusion strategies investigated in this chapter allow us to answer in the affirmative our opening questions: Girls and women do have a lot of fun when using ICTs, and promises of enjoyment do to a considerable extent make ICT more attractive for women as well as men. Hence, fun, play and glam can be very effective ingredients in initiatives to increase women's use and knowledge of the technology and thus help them to become included in the Information Society. Pleasure in technology should be available to all; it is a potential to be encouraged, not a problem to be avoided, with respect to digital inclusion. Accordingly, we would propose that, in contrast to the puritanical flavour of earlier discourses on women and computers, the language of *desire* and *attraction* should be added to the lexicon of digital inclusion – to capture just how significant fun, play and glamour are in people's motivation to engage with ICTs.

By demonstrating how girls too are having fun with ICT, the research presented here challenges the gender dualised presumption that boys play with computers while girls do not. But whilst we have seen clear shifts in people's practices, the tendency to dichotomise women's and men's computer use symbolically has proved remarkably resilient with either/or categories – toy/tool, fun/serious, right/wrong – mapped time and again onto a presumed gender dualism. Worryingly, these co-constructions often accompany stories that 'girls and women are not doing it right'. Thus the fun young men had with computers at The Gathering was perceived as more serious and more skilled than that had by young women enthusiasts. Designers of fun-based inclusion strategies as well as many users may struggle to transgress gender dichotomies and hierarchies. Fun is consequently not a quick fix to break out from gender dualisms, or to inclusion.

Nevertheless, fun complicates the gender-technology picture. Computers are both tools and toys, for both girls and boys. By destabilising the either/or of fun and ICT, the ICT options for girls and young women are extended. The diversity and complexity of ICT use and fun revealed here empirically is hugely significant, also because it profoundly challenges the social/technical dualism on which the hacker myth drew. Even if some men are more enthusiastic about the technology than the majority of women; not all men are enthusiasts and some women are. Moreover, both men and women enjoy the strongly social aspects of their professional work – a reminder that the a-social nerd was never an appropriate model for ICT occupations! Here as elsewhere in this book, we find that gender diversity rather than gender binaries more accurately captures actual ICT practices, and we may conclude that inclusion strategies work better when based on diversity rather than dichotomies.

From the study of the Norwegian secondary schools, we learnt that the effort to exorcise 'desire' from the computer was successful: computing was made boring to the pupils. But as we saw, the teenage girls and boys alike spent considerable amounts of their leisure time having fun with ICTs in various ways. This significant change has been due more to the advent of new leisure and communicative applications of ICTs than to any deliberate inclusion efforts. As a result, the early research findings pointing to a boy/toy vs. girl/tool binary have quite simply been superseded by sociotechnical developments. In the course of the domestication of ICTs for fun uses, the phenomenon of the computer as an object of desire in itself has declined, but in the process the computer has become an instrument to do desirable things – and for increasing proportions of the population. Thus, it is time to do away with the boy/toy vs. girl/tool binary. Fun and play may be important ingredients in effective inclusion strategies targeted at girls and women.

However, whilst the strategies of fun, play and glam clearly have succeeded in making more girls and women competent users of different ICTs, there is no evidence as of yet that these strategies have inspired more young women to opt for a career in ICT. Indeed, there are hints that many 'girls just want to have fun' with computers and (ironically) are uninterested in 'serious' ICT. In any case, the evidence presented here suggests that fun and enthusiasm may not be an effective strategy for attracting more women into higher education in computing and a career as a computer professional – given the distinction the studied teenagers made between fun leisure computing and boring school computing, and the perception that fun technology use is not computing or technology per se. As we saw with IT Beat, making ICT relevant to people's lives often involves moving the technology backstage and so making it invisible and almost trivial. Yet by this stroke, the strategy also moves a main inclusion message backstage.

We return to such issues in Chapter 7, which details a comprehensive inclusion initiative that aimed at recruiting and retaining more women computer science students. As we will see, the main motives of the women computer science students actually transcended the tool/toy binary. Put a bit crudely, they found computer science attractive, not because it was fun or allowed them to be useful, but because the education offered good career prospects. Desire has many dimensions.

In the next chapter, we pursue issues related to tools and toys in a different way, by focusing on strategies to provide designs of ICT products that are attractive to women. How do companies go about finding out what women desire in ICT products, and what representations of women are made use of when designing for women?

6

Inclusion by design?

Els Rommes

In Chapters 4 and 5, we observed that some products such as web magazines actually had been designed for women or particular groups of women. We found that these products seemed to be strong in providing relevance to their women users, thus addressing the important inclusion need of motivation. In this chapter, we analyse why ICT products may be geared towards women and how gender sensitivity performs in design. Consequently, we are mainly studying actors in the ICT and new media industry and how they approach the challenge of including women either to increase their potential market or to find new outlets. To what extent do business concerns produce inclusion activities?

Much of the research on the exclusion of women from ICT (or technology more generally) identifies as a key factor the fact that the development and design of technology is dominated numerically by men. The underlying assumption has been that men design for men (Wajcman 1991, 2004). This in turn has been seen as a strong reason (apart from issues of equality) to increase the proportion of women in technological development and design work. In the following two chapters, we address efforts to get more women to study and work with ICT. Here, our main concern is to explore, based on our studies, what happens when (mostly men) designers design for women, and what scope there is for this to be effective.

At first sight, it would seem a truism to suggest that design efforts informed from thinking about gender should yield products that are more attractive and better suited to women. However, the matter is more complex than this. First, the whole process of mapping needs and translating needs into design is rarely straightforward (Ehn 1988, Williams et al. 2005). Second, to identify features of a product as 'masculine' or 'feminine' presumes that genders easily translate into physical properties of artefacts or screen images. In practice, such translations between the social and the technical rarely work and at best have local validity (Sørensen 1992; Suchman 1994; Suchman 2009). Third, although the design of a product can be understood as an effort to 'configure' the users, users have considerable freedom to domesticate products and thus re-configure intended practices and meanings (Lie and Sørensen 1996; Oudshoorn and Pinch, 2004, Berker et al. 2006).

Perhaps the most outspoken effort to design products for women and girls has been in the making computer games (Cassell and Jenkins 1998b). The most famous example, Barbie Fashion Designer, was an immediate commercial success (Castell and Bryson 1998, Jenkins and Cassell 2008). Cassell and Jenkins (1998a) argue that to some extent the movement for making computer games for girls was fuelled by the entrepreneurial feminism of the 1980s and 1990s. Some women entrepreneurs started their enterprises with the explicit goal of furthering the political and social goals of the feminist community. Yet many of the computer games which came out of this movement seem to have employed stereotypical representations of femininity as a point of departure (Cassel and Jenkins 1998b, Kafai et al. 2008, see also Spilker and Sørensen 2002). How should we think about inclusion where it is achieved by invoking stereotypes that may serve to reproduce traditional gender relations?

These observations indicate that promoting gender sensitivity in the design of products can be what Donna Haraway (1991) describes as a trickster. It may not work in the anticipated manner. A common strategy is to try to make products more relevant and attractive to girls or women by introducing designs that are different from those that seem most attractive to men (Cassell and Jenkins 1998b). However, when this is achieved at the cost of reproducing stereotypical representations of girls or women, the gender politics of the design efforts become more problematic. From our perspective, inclusion is about changing not only

ICT products but also gender relations and representations. Are there, or could there be, design practices that reflect this perspective?

Our focus in this chapter is mainly on the practices of those companies which have an explicit concern with gender issues and which reflect upon design strategies intended to reach a large audience or market segment of women. Such companies are the most interesting for our purposes, since they actually are linking design strategies and inclusion of women. We studied four web magazine presses, one producer of consumer electronics, five computer game companies and one mobile phone company – in total eleven enterprises. Our analysis is in three parts. First, we analyse how these companies understood the challenges of including more women users, and their motives for making inclusion efforts. Second, we study how the companies actually took gender into account. To what extent did they work with stereotypes, to what extent did they make efforts to effect changes in gender relations? And third, we investigate the methods adopted by the companies to try to take account of the wishes and interests potential girl or women users in their design efforts and what kind of design strategies they pursued.

In Chapter 1, we introduced two political strategies for achieving equal opportunities for men and women: category politics and gender mainstreaming. In this chapter, we consider whether the distinction between these two strategies makes sense in commercial environments. In this setting, we suggest, category politics means designing explicitly for women, while gender mainstreaming implies an effort to provide designs capable of including (some) women as well as men.

Inclusion of women: Commercial motives?

We observed instances of entrepreneurial feminism in two of the web magazines for women that we studied but in none of the other companies. While all were concerned with girls or women because they saw this as market opportunities of some kind, their perception of the opportunities varied. The four web magazines were thoroughly analysed in Chapter 4. As we showed there, these magazines targeted women and made efforts to get women users mainly through content but also by offering tailored services, such as instruction in the use of computers and the Internet, or by providing space for interactivity, such as chat rooms.

The multinational Philips, founded in 1891, is one of the largest producers of electronic consumer goods and one of the main actors in multimedia in the Netherlands (Lente 1999: 309). In 1993, employees of the New Business Creation group in Philips Research decided to develop a product for the electronic games market [Rommes et al. 2003], and decided to look for a niche in the area of electronic devices for girls. The main driver for this was the strategy of Philips to try to obtain at least a significant, if not the major, share of any market. One promising direction appeared to be the electronic video games market that was at fever pitch. The market for electronic games for girls had just been discovered, with the launch of Mattel's 'Barbie interactive line' A connection was made between the potential niche of a girls' games market and the fact that Philips at that point had the lead in developing Liquid Crystal Display screens on which users can write with a special pencil, and in developing infrared-based communication technology. This technological development facilitated a communication device, which the Philips employees perceived to be particularly suitable for girls. The product was called KidCom.

Because the gender of the target user differed considerably from Philips' more common target user groups, the composition of the design team, the design methodology and the user representations were given more attention than in other Philips projects (see, e.g., Oudshoorn et al. 2004). The KidCom project was intended as an example of how to do design 'the right way', as they did not want to take any risks in entering this new market segment. Children participants influenced the development trajectory of the project and the type of product that was developed to a much larger extent than was common in Philips.

Moreover, gender was of major importance in the composition of the KidCom team. For instance, the number of women that were involved was equal to the number of men involved. The jobs, for which women were hired, were directly related to knowledge about girls, such as concept development and design of the product, and the making of the marketing campaigns. Several kinds of designers and marketers were intentionally included, in addition to engineers, with the goal of ensuring that enough attention was paid to the users. KidCom was developed as an electronic portable device for girls and its marketing would be particularly directed at girls, but it was to be presented as a tool with communication options for 'kids' because it was conventional wisdom in the video games market that 'boys would never play games intended for girls'. Although the KidCom was received with great enthusiasm by girls, for internal reasons the company never put it on the market.

The games of the other Dutch firm we studied, a fairly small company founded in the late 1990s by three graduates of interaction design, did make it to the market and were very successful [Stewart 2003]. The three owners of the company considered themselves to be designers and artists who created and sold creative interactive computer-based products to publishers and other intermediaries in the market. Rather than developing products for a mass market, this firm worked on commission for museums, schools and educational organisations. As one of the designers in this company said: 'We are in the edutainment business, not the games market. We notice it [the games business] is a whole different ball game. The boyish style, the whole culture is totally different to how we are and what we do'. Accordingly, they produced a series of 'girl games', entertainment CD ROMs for young teenage girls on behalf of a major general store chain which sold a range of products aimed at girls.

The games made by this company were developed by a team of three: one woman graphic designer, one woman designing concept and graphics, and one man who did the programming. This was similar to the sex composition of the development team of Philips, graphical and interaction designers had at least as much to say as programmers. We were told that the style of the games was very much to be based on what the designers wanted to make. 'The first thing we noticed that it gave opportunity to play this game in a certain way which was different from the way that boys played it, which was clearly noticed in tests'. They claimed to have discovered a girls' market by chance. By testing their finished games on children, they discovered that their particular style was popular amongst girls.

The four Norwegian computer game companies we studied tended to pursue different strategies to include more girls and women as users. Moreover, the interviewees all expressed greater ambiguity about gender issues [Gansmo et al. 2003]. To begin with, the companies were all sceptical about designing games specifically for women – partly because they doubted that the market was large enough, but above all because the designers felt unsure about how to transform 'feminine' tastes into 'good' games. Consequently, the companies opted to try to make their games attractive to women as well as men, rather than to target women. Interviewees in two of the companies said that their main inclination was to stick to the market segment they knew, which consisted of young men, but that they wanted to cater to the tastes of some women also. However, the third company, Caprino, had chosen to pursue what could be considered a family market, which meant they had to try to accommodate the tastes of girls as well as boys. Interviewees in the fourth, Artplant, told us that they tried to design quality games for smaller market segments, based on television series, magazines or book series. They had recently made a game for a horse club magazine, which in practice had girls as its main target audience. However, when girls became their target audience, it followed from the content of the game, not because they particularly chose to target girls.

The main strategy of inclusion, as it was described to us by the interviewees from the Norwegian game companies, was to try to target girls and women as well as boys and men by designing 'quality' games with 'good' and well-known stories. For example, some designed games that combined role playing and strategy, believed to be important to women, with action elements, believed to be important to men. In addition, some of the companies thought that attractive packing was an important means to recruit more women customers.

The other business we studies was a research laboratory (TILAB) subsidiary of Telecom Italia [Fortunati and Mangelli 2003]. When we did our interviews, TILAB was working with what was then a new generation of mobile phone technology, the UMTS. Part of the laboratory was engaged in researching users and consisted of four women and two men. They were concerned with the inclusion of women, although their efforts on this met with difficulties stemming from lack of resources (most products were given a maximum of three weeks of research time on users in the final phase of product design), unclear organisational responsibility for such issues and an emphasis on purchasing power of the customers by the producers of the products and the researchers at TILAB, which led them to neglect needs and wishes of non-users and dropouts. The examples of inclusion efforts provided were technological, like a new kind of platform and vocal portals, and stemmed from dichotomous ways of thinking about men and women, e.g. that vocal portals be particularly suitable for women because of their oral quality and immediateness.

Why include women?

The most common way of thinking about inclusion observed across all but the Norwegian games companies we studied resembles category politics. Thus, the four web magazines had identified women or 'modern women' as their main target audience. Also, Philips with KidCom and the Dutch computer game company distinguished between products for girls and for boys. At TILAB, they thought about inclusion in a similar way, even if they did not in the end add designs specifically for women users.

What makes (some) companies choose women as a specific target group, or not? What factors in the design context may be influential in deciding whether a design team takes gender differences into account or addresses gender inequalities with their product? We might expect most companies to be less interested in normative or ideological aspects of gender inclusion than in seeing women as a potential new market or as an extension of an existing market. Having said this, there is plenty of evidence of companies failing to even consider women as a potentially profitable market. In the case of iVenus, the web magazine for women in Ireland analysed in Chapter 4: 'the fact that the target audience was female was to some extent chance. The company's off-line publishing titles are predominantly aimed at women and it was for this reason that they decided on a women's on-line magazine rather than as a result of any research into target markets' ([MacKeogh 2003e: 390–391]. And Oldenziel's sweeping review of research on gender in the history of technology, reveales that 'In many cases, the early inventors-entrepreneurs only accidentally "discovered" a new mass-market of women' (Oldenziel 2001: 136).

It may also be common that companies are reluctant to take the risks associated with targeting girls or women where these are a relatively unknown market. In the analysis of the four Norwegian game companies, Gansmo et al. [2003:137] concluded that there was 'an outspoken scepticism about designing games that are labelled as designed particularly for girls or women'. This scepticism was partially grounded in doubts about the size of the market, but it also reflected a hesitance about experimenting with tastes. The safe thing, as in most sectors, was to work from the well-known (see Walsh et al. 1992). Similar reasoning was expressed by large game producing companies such as Philips (Oudshoorn et al. 2004) and SEGA (Glos and Goldin 1998).

So what kind of companies takes the risks involved in designing for women? It seems that inclusion activities sometimes emerge by companies who have a large incentive for finding new niches because they are either relatively new in the field or developing new technologies or design styles. For example, the 'KidCom' design team had just developed a new kind of technology which they wanted to develop further and for which there was no established market yet [Rommes et al. 2003]. Similarly, the games company Stewart [2003] studied discovered more or less by accident that the specific aesthetics and interaction style they designed was particularly attractive to girls. In contrast, firms that are well established and have a comfortable segment of the market may favour evolutionary, market-led incremental design improvements based on feedback by existing customers, rather than targeting a new user group.

Some companies do nevertheless seek to include women in their ICT product designs. They may be motivated by potential profit or commissions [Fortunati 2003c, MacKeogh 2003e, Stewart 2003] but also by gender politics. Thus, one should not overlook the potential importance of political and social motives, so central to feminist entrepreneurs (Cassel and Jenkins 1998a). The presence of feminist entrepreneurs seems to have been important in some companies, not only by providing motivation but also by providing insights about the potential market that women represent and how this market might be accessed [Slooten 2003b]. Even though the manager of the Dutch women's magazine 'Libelle' stated that Libelle did not work from an emancipatory or feminist perspective, the woman who was hired to design the website did identify herself as a feminist with the main goal to 'make the Internet more accessible for women and to develop opportunities for women' [Slooten 2003b: 167]. Her contribution was vital in making the Libelle site successful in introducing many women to the Internet. Such findings are in line with a large-scale survey of successful, innovative companies, in which Walsh et al. (1992: 165) found that 'crucial factors for success appear to be the use of a dedicated project manager and the support at senior management level of a 'product champion'.

In sum, four features seem important when companies or designers become aware of a potential market of ICT products for women. First, smaller companies in search of a niche market and with sufficient resources to develop a new product may find (a subcategory of) women to be an interesting audience. Second, a company may be more motivated to develop a product for women if they have just developed a new technology or a new way of using it (like the interaction style and aesthetics described by Stewart [2003]) for which no market yet has been defined; a form of 'technology push'. Third, companies may be made conscious of the potential markets girls and women represent through market research or engagement of employees; a form of 'market pull' may be observed. Fourth, presence of feminist entrepreneurs implies an emphasis on detecting women or girls as target group and clarifying related inclusion needs.

Design practices and inclusion methods

When designers make a new product or remake an existing one, they draw on some representation of the intended users, on images of the presumed target group. This may happen more or less consciously, but one cannot really design a product without some idea about how it is going to be used, by whom and for what purpose (Akrich 1995). In the eleven companies we studied, we observed four main practices with respect to the construction of user representations, each with different implications for the way gender issues were approached:

- Designing from gender stereotypes
- I methodology, where designers are allowed to draw on personal tastes or experiences, seeing themselves as typical users.
- Reflexive I methodology, where companies deliberately chose to involve women designers.
- Participatory design, where potential users were directly involved in the design process.

In the following, we describe and discuss these practices in greater detail. Of course, these four approaches may be used interchangeably, even in the same project.

Designing from stereotypes

A lot of thinking about user representations appears to be embodied, in the sense that designers immediately invoke stereotypes to describe what users want or need. In the companies we studied, it was clear that many designers who aimed to include women as end-users of their products saw it as important to define in what ways women or girls 'are' different from men or boys. Many were convinced, for example, that some features just 'belong' to women's or girls' products 'like fashion and beauty' – as we saw in several of the web magazines (see Chapter 4). How are such stereotypes chosen and invoked?

In the KidCom project, the designers started with a literature study on gender differences, using books and reports that have in fact been extensively criticised by feminists for the way they dichotomise and naturalise perceived 'biological' differences between men and women [Rommes et al. 2003]. In most of our studies, however, we found a more impressionistic approach. For example, often popular print magazines aimed at girls or women were used as a source of inspiration [Gansmo et al. 2003, Stewart 2003]. As a result, the appearance of these products tended to have aesthetics similar to the girls magazines that inspired them, with lots of colour (especially pink!), cartoon images, and non-rectangular windows [Stewart 2003, Rommes et al. 2003]. Editors of web magazines were definitely inspired by the ordinary paper magazines for women, but also by the few web magazines for women that were around at the time the product was developed [Fortunati 2003c, Hestflått 2003]. Some designers also mentioned that they had looked at girls' books, or watched girls' favourite television programmes.

Regardless of the source, the most common way of constructing an image of potential girl or woman users of a product was by thinking from stereotypes. To some extent, this reflected a felt need to construct an image of their future users as 'different' from the audience they had targeted before. Hence, no matter who actually were thought to be the precise target group – adult women or girls, highly educated women or women living in disadvantaged areas – the same general beliefs about what women want tended to resurface. Women were represented as neither interested nor skilled in technology and as preferring 'user-friendly' (i.e. simplified) interfaces. Another widespread perception was that women were not into technology for fun but to gain from useful applications, without flashy pictures and such like. Contrary to our findings about the importance of having fun with computers and the Internet (Chapter 5), it was widely presumed that women saw technology mainly as a tool [Stewart 2003, Rommes et al. 2003].

In nearly all the companies we studied, such stereotypical ideas about how women 'are' or what they 'like', were important in guiding design decisions, if not exclusively so. Arguably, there is a positive and a negative side to designers' interest in how girls 'are' different from boys. On the positive side, a focus on 'what girls want' may serve to strengthen and give value to skills and preferences considered feminine (Cassell and Jenkins 1998: 21). Moreover, in several cases stereotypically feminine elements were used to attract both girls and boys. Norwegian game companies made efforts to design games they thought could interest a broader target group of girls and boys or women and men. To do so, they drew on the stereotype approach to include in the game elements they assumed girls or women would like [Gansmo et al. 2003]. The Dutch company made games 'for girls' that could be used by boys as well [Stewart 2003]. Similarly, KidCom was designed by making use of the stereotype approach and was originally intended and designed for girls, but was targeted to both girls and boys [Rommes et al. 2003]. So even games designed on the basis of stereotypes have potential gender transgressing and transforming effects

On the more negative side, by developing a product based on 'typical girls' interests', designers run the risk of reinforcing and re-inscribing perceived gender differences rather then challenging gender inequalities through efforts at transforming gender. In other words, the stereotype approach may easily give rise to the design of products that are specific, e.g. pink websites with utilitarian aims focusing solely on topics presumed to be of interest to women, or even worse, gender stereotypical, if such sites are explicitly addressed to women (see Rommes et al. 2010 for the different ways in which games can reinforce and re-inscribe gender stereotypes). Gender specific and especially stereotypical games may reinforce traditional gender practices and divisions of tasks (Kafai et al. 2008; Henning et al. 2009; Corneliussen and Retberg 2008; Taylor 2006). Moreover, such designs run the risk of not being commercially viable. Most of the women gamers Kerr interviewed were as interested in puzzles with a feminine connotation as in masculine games such as racing and shooting games [Kerr 2003]. Similarly, Rommes et al. found that the girl gamers Philips tested strongly preferred black and dark colours over pink [Rommes et al. 2003]. Designing from stereotypes should, at the very least, be considered a potentially risky business.

The I methodology

One of the most commonly used techniques is to make design decisions on the basis of designers' own preferences. This is frequently referred to as the I methodology, emphasising the subjective aspect of formulating user requirements (Akrich 1995, European Commission 1998, Oudshoorn et al. 2004). The I methodology has been considered problematic in the context of gender inclusive design because most designers are men. Designers also tend to have a specific, 'insider' relationship with their technology, which makes their worldview different from that of many users of their product (Rommes 2002b). Also, the label 'I methodology' is meant as an ironic comment on the fact that the subjectivity present in the proposal of user requirements is not made explicit or an issue for reflection (Berg 1996).

In most of the eleven companies we studied, some form of I methodology was practiced. However, there were modifying features when all or parts of the target audience clearly differed from the 'I'. This was particularly evident in the making of games and toys for children. The designers in the Dutch game company relied, as we have seen, mostly on the I methodology during the development of the product [Stewart 2003]. With KidCom, designers also used other inputs [Rommes et al. 2003]. And when Artplant worked to make a game with a focus on horses, they read the magazine that commissioned the game as well as doing field trips to stables and talking to girls [Gansmo et al. 2003]. It became more complicated when designers saw themselves as a potential part of the target audience, if not necessarily a typical representative.

When interviewing in the Norwegian company Funcom, it was clear that some form of I methodology was at work, for example, with respect to decisions of what was a 'good' story from which a game could be made [Gansmo et al. 2003].

The concern for gender inclusion made several companies use what we consider to be an adapted version of the I methodology by explicitly making choices with respect to the sex composition of design teams or by stating up front that women ought to know what women want and need. This, we call reflexive I methodology. What kind of design practices did this refer to?

Reflexive I methodology

In feminist methodology, Haraway (1988) has called for reflexive approaches to science and engineering, noting that such work needs to be situated to be made understandable and that all perspectives are partial. Clearly, by adhering to these tenets, the I methodology may be used in a reflexive way. This requires that designers be aware of and take into account the ways in which their particular perspectives and situation make them similar to and different from the end-users of their products. Of course, relying on stereotypes is one way of dealing with the situatedness of designers in so much as it acknowledges that they are different from the intended users. However, to qualify as reflexive I methodology, designers need to pursue in much more depth the thinking and reflection about their roles.

What happened in many of the companies we studied was that they asked women to join the design team for a product aimed at women. For example, the project manager of KidCom justified the inclusion of women in the team by suggesting that women participants 'would surely know what seven-year-old girls like and prefer. They have been there themselves' [Rommes et al. 2003; see also MacKeogh 2003a, e]. In the Norwegian game company, Caprino, we were told that 'the only female character [...] has been made a bit more visible in the game because the woman developer thought it was important' [Gansmo et al. 2003]. With a slightly different emphasis, the editor of the Norwegian web magazine for women, Femme, said she saw herself and her closest women collaborators as representative of their target audience, 'modern women' [Hestflått 2003]. Similarly, in the web magazines we studied, content choices were made by the editors, who were women. As indicated above, they made their choices mostly from their own preferences, based on some form of introspection. In all of these instances, it was assumed that women are necessarily similar to their target audiences of girls or women, and understand these potential users better than men designers would. In effect, they used the reflexive I-methodology in an essentialist way.

This said, even where women were involved positively in the design, their influence was often limited. Most lay-out and navigational choices for the web magazines, for example, were made by web designers, who tended to be men. They considered themselves to be experts on how to reach specific groups: 'It is up to us to suggest services, applications that can build ... that can give substance to their idea', including about 'how to reach women'. Their lay-out and navigational choices were mostly based on the stereotype approach. In the case of the Irish web magazine, iVenus, women had less influence with respect to design because they were not placed in responsible positions and had no opportunity to make direct inputs into the development of the site [MacKeogh 2003e]. Hence, if women are introduced in the design process as a way of introducing a reflexive I methodology, their position with the organisational hierarchy and the division of labour will have a major bearing on how much they are able to influence the ultimate design.

There remains the problematic issue of the implied essentialism in this version of the reflexive I methodology, the belief that every woman is representative of women (or girls) in general. As indicated in the discussion about the I methodology, even in cases where women designers are similar to the potential users – e.g., in terms of age, class, ethnicity and interests – the mere fact that they are situated as part of a design team, have access to the latest technology, and are interested and skilled in the use of technology, makes their relation with technology different than that of most users. So, both versions of the I methodology need to be used with care. Moreover, because of the essentialist tendency inherent in the observed use of reflexive I methodology, it is likely to lead to gender traditional products, as some of the web magazines and computer games exemplify. This danger could be avoided if designers' reflexivity were based on efforts to find out about the complexity and diversity of actual people, their practices, identities and the like, rather than on essentialist assumptions. One way of doing this is to access expert knowledge about gender, so that design teams can learn from previous research on the subject, and use this knowledge to identify relevant considerations and refine their own understandings. Another way to find out what girl and women end users is to engage in participatory and interactive design techniques, to which we now turn.

User testing and participatory design

As we have seen, both the stereotype approach and the reflexive I methodology in the form of deliberately employing women designers assume that there are differences between men and women and, consequently, risk reproducing these differences through the design of quite gender traditional products. Arguably, the most important alternative is to use of techniques such as the testing of products on potential users or participatory design. These techniques – which have sometimes been called 'feminist' strategies (Balka 2005, Cassell 1998, Greenbaum 1991, Suchman 1991) – allow potential end-users some direct influence upon design, with the result that design does not rely solely on the ideas and beliefs of designers.

Several of the companies we studied invested time and energy in identifying representative end-users and examining their preferences. The designers in the Norwegian company Artplant, who were in the process of developing a game for horse enthusiasts (i.e., girls) went to a stable, talked with girls there and asked them what they would like in a game on the theme. In a different vein, another Norwegian company, Funcom, made early versions of their new on-line game available to users so they could try out these versions and identify weaknesses and errors – what adds to a kind of comprehensive user testing [Gansmo et al. 2003]. The interviewees from the Dutch game company told they tested a game for girls at a school [Stewart 2003]. And the development of KidCom involved a comprehensive exploration of user requirements as well as user testing with children [Rommes et al. 2003].

Such user involvement in the design process seems to be the exception rather than the rule (European Commission 1998, Haddon and Paul 2001, Offenbeek and Koopman 1996). And involvement of potential users in early phases of the design process is especially rare.[22] In this context, it is interesting to note the fairly widespread use of testing and some kind of user participation in the companies we have been studying in various phases in the design process. This suggests that concerns to include more girls or women result in greater than normal engagement with potential users – possibly because of unfamiliarity with the target group, as one of the KidCom designers indicated. In addition, it seems clear that the Internet facilitates interaction with users and user input

22 It is so exceptional that several authors have developed design methodologies that they consider to be innovative because of the early phases and the extent to which they involve potential users (see e.g. Mackay et al. 2000 and Fullerton et al. 2006).

to design, which helped the companies we studied. New ways of user testing and getting feedback on designs have become available as a result of the 'interactive features of the web' [MacKeogh 2003e]. Examples of users reporting their preferences and problems on-line to designers were found in all four of the web magazines and in the development of computer games and toys. This feedback contributed in several instances to changes of designers' understanding of their users and in some cases influenced design decisions [Hestflått 2003, MacKeogh 2003e]. Slooten [2003b] reported that 'the reason for Libelle to pay attention to Internet and computers [on their web site], is simply because readers ask for it'. The staff of the magazine even used the feedback of the users of the website as input to the printed version of the magazine.

Although the interactive features of ICTs seem to be very effective in allowing users more influence, those users that give feedback on websites are a select group: more active, engaged and articulate then ordinary users. By focusing solely on their needs and wishes, designers run the risk of ignoring those of users who are less capable of articulating their demands, or of ignoring potential users that are not yet using the product. This is a general problem with respect to participatory design strategies: how to achieve a reasonable degree of representativeness? We saw in several web magazines that employees or people that 'happen to be around the office' were used to test the site, causing similar problems as those we described with the (reflexive) I methodology. The Italian mobile phones were tested in TILab, an official testing laboratory, which created 'research professionals' that no longer seem representative of the end-users [Fortunati and Manganelli 2003]. The companies we studied could perhaps have been more concerned about these limitations. Again the challenge is to find out about the multitude of actual and potential users, to become more sensitive to the diversity of wishes, skills and preferences, which could enable their designs to include a wider range of users.

Gender inclusive products through gender inclusive design?

We began this chapter by asking if business concerns can produce inclusion activities. Our findings are positive in this respect. At least some companies in the areas of ICT and new media are interested in attracting girls and women as customers, and there is no doubt that including such audiences offers interesting business opportunities. These opportunities seem to be most commonly pursued by fairly small companies or startups, but not exclusively. Other evidence suggests that large companies are getting more interested and buying small companies with a portfolio of products directed at girls or women (Kafai et al. 2008). In addition, we have noted effects resulting from new technological options (technology push) as well as concerns about market expansion. And we have seen how personal engagement, such as that found among feminist entrepreneurs, may also be a driver.

However, the translation of inclusion opportunities into successful products is not straightforward. Most companies have other interests and stakeholders, which need to be taken into account and which may conflict with or constrain such developments. For example, the interests of shareholders may supersede those of potential women end users (Oudshoorn et al. 2004). Web magazines need to consider advertisers' interests – probably one of the main reasons why women's magazines pay so much attention to beauty and fashion [Fortunati 2003c]. We also found that designers of games for children were very concerned about what parents would want to buy for their children [Rommes et al. 2003, Stewart 2003]. KidCom was given a pinkish colour to indulge parents, even though girl testers had clearly indicated that they did not like pink. Most of the games companies reported that they were dependent on publishers and distributors, sometimes also governmental agencies, and therefore had to respond to their ideas about what should be put on the market [Stewart 2003, Gansmo et al. 2003a] (see also Rommes et al. 2010).

Four main strategies for building representations of potential girl and women audiences were in use among the eleven companies we studied – stereotyping, I methodology, reflexive I methodology and user testing/participatory methods. The fact that all of the companies combined two or more of these approaches indicates that they were uncertain how to acquire the knowledge they needed about girls' and women's requirements. It also indicates that their thinking about design as a gender inclusion strategy was under construction: for all of them, this was a fairly new issue.

Each of the strategies identified here has advantages and disadvantages. The first, relying on stereotypes, represents a revaluation of values and interests usually seen as feminine, but at the same time risks reinforcing and re-inscribing the stereotypes, particularly where designs are based on presumed gender binaries. The I methodology is problematic to use if you design for users who are clearly different from yourself. It may also result in the reproduction of designs some women find unattractive because they appear masculine. However, in several cases we observed a revised version of this strategy, which we have called a reflexive I methodology. This involved the deliberate employment of women designers, based on the idea that women know what women want. Although this did, in some cases, lead to some more attention to user needs and wishes, and it definitely created more ICT jobs for women, the obvious problem of this strategy is its implied essentialism. The fourth strategy involved the application of direct user involvement techniques, for example, surveys, testing and feedback. This facilitates broader and more diverse inputs to design, but the methodology itself gives no guarantee that the full spectrum or actual gender diversity is taken on board.

Given the focus of this book on inclusion in the Information Society, it is interesting to note that the Internet emerges as a facilitator of – mediated – user involvement. Companies working with ICT and new media may have better opportunities to involve users over a longer period of time than many other businesses. As Stewart and Williams (2005) remind us, design involves complex learning processes, including trial and error. It seems clear to us that gender sensitive and gender inclusive designs need to be considered as evolving practices. It would be very interesting to study companies developing products that seek to be more gender inclusive for a sufficiently long period to be able to assess how the involvement of girl and women audiences might help companies to improve their designs so as to better meet the needs and wishes of these audiences.

Does attention to gender in the design process produce gender inclusive products? To answer this question we need to consider what kinds of products have resulted. In the companies we have studied, the most common design could be described as gender traditional, meaning that the product is made attractive by catering or appealing to traditional perceptions of girls' or women's interests, often in a singular form, as in 'the interest of woman'. This kind of design is usually produced through stereotyping or through a reflexive I methodology based on the idea that women designers are representative of women in general. It is embedded in a thinking about audiences that invites gender binaries. And it tends to invite the design of products specifically for (girls or) women – a kind of category politics.

We see the counterpart to gender traditional design as efforts to make what we are calling gender transgressive products, meaning products that somehow transcend or challenge gender binaries. Alternatively, Light (2011) has suggested the term 'queering' or 'troubling' design to identify designs in which men and women are given the space to define themselves. Somewhat more broadly, Rode (2011) suggests the term 'gender sensitive design' to mean designs that support feminist goals. Arguably, none of the products made by the companies we studied satisfy these criteria fully. Some of the web magazines such as Libelle and iVenus made efforts in this direction. The designers of KidCom could be interpreted as having struggled with being both traditional and transgressive, as did the Norwegian game companies. Significantly, the latter companies seemed to pursue a version of gender mainstreaming in design by claiming to cater for the interest of both men and women gamers. They also tried to think from a non-binary understanding of gender by differing between 'normal men', hardcore gaming men, women gamers and women not interested in computer games, to argue that their target audience consisted of 'normal men' and women gamers.

Following this line of thinking, we might recast the question about the relationship between gender sensitive design and gender inclusive products to ask: What kind of design methodologies may lead to gender transgressive products (see also a special issue on feminist design methodologies of *Science, Technology, & Human Values* (forthcoming 2012))? Sengers et al. (2005) proposed the use of 'reflective design methodologies' to make unconscious values in the design process visible and available for conscious choice. Alhutter (forthcoming) presents a method for making implicit values explicit. Although the presence of feminist values amongst designers may indeed lead to more gender transgressive games (see Rommes et al. 2010), making values visible does not guarantee that designers actually will choose feminist values and goals.

Overall, we see it as very important to promote deeper versions of gender sensitive design, as a way to include more girls and women into the Information Society. As we have so frequently observed, if 'gender sensitive design' is interpreted in simplistic and essentialist ways, there is no guarantee this will lead to gender inclusive products – precisely because of the trickster quality (Haraway 1991) of such design efforts. In particular, the ease with which gender stereotypes and binaries enter into design requirements serve as a warning of the risks involved.

7

Getting women into computer science

Vivian A. Lagesen

Analysing exclusion and inclusion: Women in ICT

The previous chapter observed how some companies consciously used women designers to design ICT products for women. While this strategy has substantial weaknesses, it also faces the challenge of actually having access to women ICT experts. In this and the next chapter we analyse efforts to include more women as producers of ICTs and new media: that is women who take a higher education in computer science and professional women in the ICT industry. While the gender gap with respect to use of computers and other new media is decreasing (see Chapter 1), it remains high and rising within higher ICT education and the ICT workforce (Charles and Bradley 2006, Bartol and Aspray 2006). What kinds of inclusion strategies have been used to remedy this situation, and how do they work? The inclusion of women into the ICT professions also raises critical issues with respect to the gender-technology relationship and the way it may be perceived and performed.

There is a large body of research concerned with women in higher education in computer science (see, e.g., the overviews in Barker and Aspray 2006 and Lagesen 2006). Most of this research has pursued what we in this book have called an exclusion agenda, with less concern for processes of inclusion. This chapter focuses primarily on inclusion issues. But, since inclusion efforts tend to be informed by exclusion-oriented research, we first need to discuss some of the main results from this research as a backdrop of the analysis.

Women comprise a small minority of the students in computer science and engineering in most countries (see Chapter 1). This situation – 'the woman problem in computer science' (Lagesen 2005) – began to become visible in the early 1980s through observations that young women took less interest in computers than young men (e.g., Dambrot et al. 1985). The main issue in studies of this problem has been the question 'why so few?' Many accounts have been offered (Ahuja 2002, Cronin and Rogers 1999, Cohoon and Aspray 2006a, Dryburgh 2000, Littleton and Hoyes 2002, Wilson 2003, Sing et al. 2007), with striking metaphors like 'the incredible shrinking or leaking pipeline' or 'the Silicon Ceiling'.

The main line of reasoning has been through the use of so-called deficit models. They provide explanations by focusing on shortfalls, deficiencies and insufficiencies. We may summarise previous research as dealing with three different sorts of such models: (1) deficits in women, (2) deficits in computer science, and (3) deficits in the image and culture of computer science.

Examples of research focusing on presumed deficits in women include claims that women have weaker computer knowledge and lesser computer experience. Such claims tend to accompany the view that women are more computer anxious than men (Brosnan 1998), that they are less confident about their computer skills (Beyer et al. 2004, Borge et al. 1980, Maccoby and Jacklin 1974, Varma 2002), that they lack self-efficacy in relation to computer science (Durndell et al. 2000, Galpin et al. 2003), or that they are alienated or just not very interested in computers or computer science (Rasmussen 1997, Siann 1997, Symonds 2000). Such models of women's deficit are grounded in comparisons with men's use of computers. This is presented as a standard that women fail to reach (Kramer and Lehman 1990). From this perspective, getting women into the discipline of computer science requires 'compensatory strategies', such as providing better information, remedial training and special encouragement for girls and women to enter the field (e.g., Cronin and Roger 1999, Henwood 2000).

Many studies have been concerned with insufficiencies in educational practices and in the content of computer science as obstacles to retaining women. Thus, improving the quality of computer science teaching has been seen as a good way to recruit and retain women students

(Cohoon 2002, Lagesen 2005, Margolis and Fisher 2002, Roberts et al. 2002). For example, programming courses have been found to be perceived by women students as repetitive, decontextualised, and meaningless, arguably because programming exercises tend to be abstract and without relationship to real-world problems (Balcita et al. 2002, Clegg and Trayhurn 1999, Countryman et al. 2002, Margolis and Fisher 2002, Weinberger 2004, Wilson 2003)).

Some have proposed that the content of computer science is in itself 'masculine', since the discipline of computer science has emerged from and is associated with institutions and practices usually perceived as 'masculine', in particular mathematics (Dambrot et al. 1985, Gressard and Lloyd 1987, Mahony and van Toen 1990). It is argued that when mathematics appears to be unattractive to women, computer science will be too (Kvande and Rasmussen 1989, Mahony and Van Toen 1990, Stepulevage and Plumeridge 1998). A parallel argument is that computer science grew out of military efforts and therefore is influenced by a 'masculine' way of thinking (Mörtberg 1987, Edwards 1990). Thus, it could be important to situate higher education in computing in faculties of social science or humanities instead of science and engineering (Kvande and Rasmussen 1989).

Many scholars have explored the assumption that computer science has a 'masculine' image which is a turn-off for women. For example, computer science students have been seen as unsociable and nerdy compared to students in other disciplines, with the result that computer science has been perceived as providing an unwelcoming classroom or workplace environment for women (Beyer et al. 2004, Jepson and Perl 2002, Weinberger 2004, West and Ross 2002). Many studies have focused on the negative impact upon girls and young women of 'a new form of masculinity' related to computers, described by reference to hackers, nerds or geeks (Kiesler et al. 1985, Berg 2000, Margolis and Fisher 2002, Rasmussen 1997, Rasmussen and Håpnes 1991). Weizenbaum (1975:116) famously described hackers at MIT as "bright young men of dishevelled appearance, often with sunken glowing eyes, can be sitting at computer consoles [...] their rumpled clothes, unwashed and unshaven faces, uncombed hair all testify that they are oblivious to their bodies and to the world in which they move".

This is in concert with the familiar view of hackers as weird, a-social but also brilliant and devoted creatures. Turkle (1984: 216) observed how the hacker culture was a 'men's world' and apparently alienating for women; "though hackers would deny that theirs is a macho culture, the preoccupation with winning and of subjecting oneself to increasingly violent tests makes their world peculiarly male in spirit, peculiarly unfriendly to women". Later, she also proposed that women are 'computer reticent' because the hacker stereotype suggests that intimacy with the machine cannot be combined with intimacy with people – something that appears inimical to conventional femininities (Turkle 1988). Faulkner (2000a) points out that this argument rests on the conventional gendering off a technical/social dualism, and the presumption that being 'technical' is mutually exclusive with being 'social'.

The prominence of the hacker figure in many studies of women and computer science has served an important role, not least in shifting the 'blame' from women. Ironically, however, it may also have contributed to the reproduction and reinforcement of unhelpful stereotypes about computer science and computer scientists (Gansmo et al. 2003a, b) and about women and men (Faulkner 2000a). For this reason, it may be necessary 'to forget the hacker' (see also Chapter 5).

As well as being off-putting to would-be woman entrants, the image of computing is also seen, within this third deficit approach, as impacting negatively on women's sense of their abilities. Also, many have emphasised the need to remedy the lack of women role models and thus increasing the visibility of women in the programme (Richardson and Kavanagh 1997, Roberts et al. 2002, Townsend 2002). A common observation is that computer science is perceived to be more difficult by women than by men, and that women perceive themselves (inaccurately) to be less competent in computer science than men (Beyer 1990, 1999, Corneliussen 2002, Nordli 2003).

Finally, it has been observed that just being a minority creates problems for women in computer science (Dambrot et al. 1985, Berg 2000, Spertus 1991, Teague 2000). Minority problems include discrimination, invisibility, scepticism and sexual harassment (Berner 1975, Kanter 1977, Dambrot et al. 1985, Spertus 1991, Barriers to Equality in Academia: Women in Computer Science at MIT, 1983) as well as isolation from other women who might provide support and solidarity in coping with these problems. Faulkner (2005) highlights a key problem of being in a small minority – the 'in/visibility paradox', whereby women in engineering are simultaneously visible as women and invisible as engineers. The visibility issue represents an extra layer of work which women, and not men, must do in order to be taken seriously as engineers and to be seen as belonging to their professional community and environment (see also Tonso 1997, 1999, 2006). Thus, the importance of increasing the networking and support communities among women students has been stressed (Gabbert and Meeker 2002, Lagesen 2007).

The general problem with deficit reasoning is that it produces a greater focus on what excludes rather than what includes women into computer science. Or, to be more precise, inclusion tend to be understood as efforts to do away with deficits, for example, by addressing the 'masculine' coding of computer science (see Chapter 2 for an extended argument). Actually, fairly few studies have explicitly addressed inclusion efforts or features that may produce inclusion. Some have advocated the need to broaden the scope of computer science (Salminen-Karlsson 1999) to include social and organisational aspects of ICT, on the presumption that contextualising computer science in this way will make computer science more woman-friendly (Clegg and Trayhurn 1999, Henwood 2000, Siann 1997). In a similar vein, it is also reported that computer science programmes located in social science or humanities environments attract a much higher proportion of women than those located in science and engineering environments (Kvande and Rasmussen 1989). Wigfield and Eccles (2000) suggest that changing the curricula to make computer science more applied and oriented toward public benefit has attracted more women. Björkman et al. (1998) and Gokhale and Stier (2004) report on women's positive perception of academic environments, while Fan and Li (2005) and Lagesen (2008), analysing gender and computer science in Taiwan and Malaysia, respectively, observed no deficit problems.

An interesting example of an inclusion project is analysed by Jane Margolis and Allan Fisher (2002). They describe a comprehensive initiative to recruit and retain women in computer science at the Carnegie Mellon University in the US. This project succeeded in increasing the percentage of women entering the School of Computer Science from 7% in 1995 to 42% in 2000. The university also managed to considerably reduce the dropout rate for women, to about the same low level as for men (less than 10%). Several measures were used. Admissions policy was changed so that admission criteria did not give a strong preference to highly experienced students, who tended to be men. Educational reforms included efforts to reduce the effect of the experience gap of men and women. More attention was paid to good teaching, since failures in pedagogy or in curricular integration turned out to affect women disproportionably, and to contextualising computer science, since women students seemed to find real world uses particularly appealing. Efforts to change the hacker-like culture of the School were also undertaken, but these were not deemed very successful. The outreach to high schools in order to make teachers more conscious about the way they influenced girls' relationship with computers, and how they could motivate girls to study computer science, proved more effective.

The Margolis and Fisher study suggests the fruitfulness of using a range of inclusion instruments and the importance of long-term efforts to create positive circles of inclusion. Moreover, it shows how exclusion thinking from deficits may be combined with an overall focus on inclusion. Observations of deficits are probably important to inform inclusion efforts when integrated in an inclusion perspective. Still, there are important challenges related to the understanding of how inclusion efforts may be organised, shaped and implemented – including concerns about the time-frames, organisational support and resources needed to achieve success (Fox et al. 2009). Also, if positive circles of inclusion may be created, what does that mean with respect to the gendering of computer science?

This chapter explores these issues by drawing on a study of a successful inclusion project, initiated by Norwegian University of Science and Technology (NTNU) in Trondheim [Lagesen 2003, 2004]. This effort, called the Women and Computing Initiative (WCI), succeeded in raising the percentage of women in the computer science and engineering programme from 6 to 38 percent in one year. We begin by outlining the main features of the WCI strategy, analysing its components and how the strategy addressed both recruitment of new students and the challenges of retaining them. The conclusion discusses what features of the inclusion initiative that made it successful.

The WCI: Construction of an inclusion strategy

In the autumn of 1996, NTNU's newly elected Pro-rector took an initiative to recruit more women to the computer science programme. At this time there was a widespread and explicit concern for the low and decreasing number of women students. The percentage of women taking a degree in computer science and engineering at NTNU had dropped from 20% in 1985 to 6% in 1996. Many challenges had to be faced if they were to recruit more women students: Few women had the upper secondary school qualifications in science subjects needed to apply for the programme. Of those who had, very few women applied to the computer science programme. Those who applied and were admitted often did not accept the offer. And the dropout rate among women students was high compared to that for men students.

The Pro-rector proposed an annual quota for women students to the Department of Computer and Information Science (IDI), based on an expansion of the total number of students to be admitted. Her proposal coincided with two other initiatives: one from the Dean of the Faculty to which IDI belonged, the other from the Head of Department of IDI. Both made a case for expanding the capacity of IDI, based on the growing demand for ICT competence in industry. These initiatives were also responses to an increased attention in Norway on the general need for more people with IT education. The Dean saw this as an opportunity not to be missed and wrote a letter to the Rector, arguing that the national capacity of IT-education needed to be doubled and that NTNU ought to do its bit to achieve this goal. The Dean's and the Head of IDI's proposals fitted neatly into Pro-rector's initiative to get more women students to IDI by allowing a quota by expanding admission. The Dean later confirmed that it was relatively easy to gain support for a quota for women as part of a general move to increase student numbers. In any case, according to the Head of Department, IDI had for many years been concerned about the low number of women students. They had made some modest efforts to recruit more women by shaping information material and brochures in a way that was intended to make it more attractive to women, and by explicitly inviting women to apply. However, the new alliance opened up possibilities for more radical initiatives.

These parallel efforts were to some extent accidental, but they highlight the benefits of piggy-backing inclusion efforts aimed at women onto general efforts to recruit more students. At NTNU, this joining of interests helped to create a powerful alliance of stakeholders, which solidly anchored the WCI amongst both the top leadership of NTNU and the staff and students in IDI. The institutional commitment as well as the resources provided was substantial.

On February 20th 1997 the NTNU's Board made a resolution of an annual quota exclusively for women. The proposal was to recruit 30 extra women for the first year and 45 for the next, allowing a 10% reduction in credit points from upper secondary school exams. In April of the same year, a WCI team was established and a half-time project manager was appointed. The WCI team immediately initiated an extensive advertising campaign. They hired an advertising agency which worked closely with the WCI manager and a representative from IDI. The campaign consisted of a screen advertisement (shown during one month at all cinemas in Norway), a website, and information brochures that were sent to girls, science teachers and advisors at 326 upper secondary schools. The campaign was funded by the Confederation of Norwegian Business and Industry (NHO) and some large Norwegian companies.

During the spring of 1997, the number of women applicants doubled compared to previous years. 22 of the women were qualified according to the general admission criterion and 46 were qualified with a 10% lower qualification. (This entry level was still higher in other engineering programmes at NTNU.) By the closing date, 52 women and 86 men had accepted their offers to become first year students at IDI. As already mentioned, this meant that the proportion of women among first year students increased from 6% to 38% in one year. This was a very high percentage of women in a European context.

In the years following this initial success, the NTNU made the WCI permanent, recognising that inclusion efforts were needed on a long-term basis. The main reason for this was that when they tried to scale down the initiative, they immediately experienced a drop in the number of women applicants. Thus, the WCI has been allowed to develop an inclusion strategy over time that consisted of a lot of different actions. In the next section we describe actions that were given priority, and the underlying understandings of women's inclusion into computer science and engineering.

Visibility and diversity: Actions within the WCI

The most striking feature of the WCI lies in the range and scope of the initiative. It included diverse actions that targeted potential women students in different phases of the education cycle. This was based on an understanding that a long-term persuasion effort was needed, and that it needed to tackle recruitment, socialisation and retention issues. The strategic aims sought to meet the challenges at four stages:

- Establishing an interest among upper secondary school girls in science in general and computer science in particular.
- Persuading young women leaving upper secondary school that it would be a good option to apply for the computer science programme.
- Persuading women admitted to the computer science programme to accept the offer.
- Providing activities to help avoid women dropping out from the computer science programme.

Each of these stages involved several actions and inclusion instruments.

Actions targeted at school girls

Some actions aimed at informing and motivating girls in upper secondary schools to choose science subjects. These included collaboration between computer science students and an upper secondary school in Trondheim, where the students would help to tutor pupils in relevant classes. A counselling and motivation course in mathematics for first grade girl pupils at seven upper secondary schools in Trondheim had been running for two years before the Women and Computing Initiative started and was felt to be successful. This collaboration involved women students from IDI teaching mathematics to school girls could come in evening classes. There was also a 'visibility' initiative directed towards young people in upper secondary school and college, which aimed to make computing visible and look fun through popular lectures about computing, demonstrations of computer-related activities and 'play' laboratories where teenagers could use computers.

These activities could be interpreted as an expression of an underlying deficit-in-women model, in so much as the main challenge was to improve girls' knowledge about, and insight into the fun and excitement of, mathematics and computer science. In the language of the inclusion needs identified in Chapter 3, it contributed to the motivation and capability of the school girls to (potentially) enter computer science.

Actions aimed at getting women to apply: The quotas

Perhaps the most radical inclusion instrument used by WCI was the introduction of a quota for women students admitted to the computer science programme. The quota was controversial and, according to the Pro-rector who initiated it, the idea needed some time to mature. However, the students at IDI were mostly very positive about the quota:

> I think it is positive for both boys and girls. I think the boys are very happy that there are not just boys in the class. That would be like being back in the army. [...] the culture becomes much better (fourth year woman student)

> I find it totally okay that women have been allocated quotas. When there was six women in the class last year, that is definitely too little. They did combine the quota with a focus on women. They had kind of a [...] invited them here and did different things. So in general I think it worked very well with more women in the programme. It does have a positive impact on the student environment and those kinds of issues (fourth year man student).

These views, which were shared across the WCI, reflects a different deficit model – that of a deficit in the culture of computer science. Something was wrong, and radical efforts were needed to instigate change. In the language of inclusion needs, the quota system clearly had a direct bearing on the number of women students admitted and so increased the access of some women to computer science education. As noted earlier, this was an outcome that was widely welcome by staff and students alike in the IDI.

By contrast, students and some faculty from other programmes at the NTNU voiced strong resentment towards quotas. The students' arguments were twofold: first, it was unfair to the men who were not admitted to the benefit of women with lower grades; and second, it was unfair to the women students because they would or at least could be seen as B-students. Such arguments are commonly mobilised in opposition to anything seen as 'positive discrimination' (e.g., Lee et al. 2010). In addition, the professors outside IDI who were against the policy, argued that it would 'steal' much coveted women students from their own engineering programmes.

The debate was heated, and student representatives on the University Board threatened to take the issue to court. The WCI had to seek legal advice from the Ombud for Equal Opportunities, who told them that a 10% reduction in credit points for women from upper secondary school was acceptable. After this, the resolution went through in the Board, in spite of resistance and public criticism from the student representatives. The university leadership admitted that the process had not followed best democratic principles. However, a professor who had put much effort into the initiative viewed it like this:

> It was a demonstration of power, a pure abuse of power. They [the university leadership] prepared everything in advance, and then they faced the critique afterwards. It is an efficient way to get things done, though (Professor Y).

According to Professor Y, the resistance from people outside the department served to unify students and staff within IDI. Later, the WCI management wrote several reports giving evidence that the WCI had probably had a rub-off effect that contributed to raising the number of women in the other engineering programmes as well; and that the women admitted to IDI had very good grades – thus faulting the contention that those women would become B-students. Still, some women students complained that the B-student label was sometimes used by fellow students to put them down.

The advertising campaigns

One of the most widely visible activities of the WCI was their advertising campaigns. That they were designed by professional agencies was unusual for Norwegian universities in the 1990s. Three versions were run between 1997 and 2003. They pushed some of the WCI arguments to the extreme and so attracted a lot of attention. The underlying strategy was to make the computer science programme attractive to women by redefining the discipline. To do this, the campaigns mobilised the conventional gendering of the technical/social binary, the main message being that computer science is a 'feminine' and non-technical one, not a 'masculine' and technical one. In this way, the advertising efforts drew upon (and challenged) a deficit model of computer science as being just technical. It also drew upon and challenged a deficit model of the image of computer science and computer scientists.

Thus, in the two first campaigns, men were mainly identified as hackers and denounced for being too absorbed in narrow technological matters. Women, by contrast, were portrayed as having better social skills, being more versatile, better at communicating and having a particular interest in users rather than in the technology in itself. The message was that women make better computer experts than men, and are more in demand by the ICT industry than hacker-like men.

These advertisements represented an understanding of the underlying gender-technology problem that is captured in the 'deficit' literature on the hacker figure described earlier – that women are put off by technology but attracted by the idea of communication and working with people, and that they view themselves as less technical than men. The adverts invoked familiar gender essentialisms around the technical/social dualism, albeit with a progressive intention. The idea was to reverse the conventional value hierarchy – the notion that people concerns (read: women's skills) are subservient to technology (read: men's skills) in computing. In this way, the campaigns stereotyped women and men, whilst challenging stereotypes about technology. This is well illustrated by the front page of one of the brochures. The text here read: 'Women make circles, men make squares. NTNU wants more computer science students that make circles'.

The third campaign drew on the same understanding of the gender-technology problem but introduced a principle of complementarity rather than a reversed hierarchy. Men and women were depicted as different, but equally wanted and needed by the software industry. As the brochure put it:

> While women use approximately a thousand words more during 24 hours and wants to 'talk about it', men still prefer to do something straight away and practical when they cope with challenges. Computer Science and Communication Technology are meant for both (author's translation).

We were particular keen to find out how the target audience reacted to these messages, and their underlying understandings of gender and technology. Most of our women informants had noticed the advertisements and had been made aware of the programme through them. Many reacted positively to the message that computer science offered good career prospects. Still, the students targeted by the campaigns did not appreciate the gender essentialist messages; in general these were ignored. Some found this aspect of the campaigns catchy and entertaining, but not really appealing; some didn't remember it; and some found it stupid. The main message coming across to the women we interviewed was that 'we want women to apply!' Many remembered the campaign as a message that they were wanted and welcome. The campaign as well as the overall initiative received considerable attention by the news media, which reinforced the impression that NTNU's invitation was serious. So the advertisements had worked as a means of motivating students to join the computer science programme, but not particularly because of the messages it contained about gender and technology.

When questioned, the women students said they would prefer that the kind of gender stereotypes used by the advertisements were avoided. They wanted to be valued for their individual qualities, and they did not believe that men and women were that different:

> It should be possible to focus more on the differences between people, not depending on whether they are men or women. They should have a focus on people that care about other people, instead of women that care about people and men that care about technology (Fourth year woman student).

> It [the hacker campaign] was a bit … this is a long time ago, but I think I found it a bit naïve, and that it was a bit … oh, I don't know. Frankly, it was a bit stupid. I don't think I would have seen it and thought that computer science [was interesting] … if I never had considered it before. I don't think so (Fourth year woman student).

The women computer science students we interviewed had a genuine interest in science subjects and some in technology too, when they applied to be admitted to the programme. Their understanding of computer science tended to be quite instrumental and pragmatic. Their alternative choice of university education would in most cases have been other engineering or science programmes. Thus, the assumption underlying these campaigns – that 'women' are not really interested in science and technology subjects but want to work with people – made these campaign run the risk of not being appealing to the arguably most motivated and qualified women.

Only one informant remembered being attracted by the effort to redefine the gendered meaning of computer science. She had applied because she was fascinated by the message of women having better communication skills and being people-oriented. In the event, however, she became highly dissatisfied with the programme and dropped out because it failed to meet her expectations of being more about people than machines. Clearly, by taking a very extreme stance on the deficit model of computer science as too technical, the campaigns misrepresented computing and misled women!

Other measures aimed at getting women to apply

In addition to the main efforts outlined above, the WCI deployed an impressive array of actions, some of them very imaginative. One was a mass dispatch of SMS-messages to a thousand women between 18 and 19 who were asked to press the OK-button if they wanted more information about the WCI. Another was to contact local newspapers before the 'Women's Day' (see below), to notify them that a girl from their region had been admitted to the computer science programme and was going to participate. She encouraged the newspapers to interview the girls and hired a photographer to cover the event. This latter action, like the Women's Day itself, may be viewed as a visibility strategy, to make women noticeable and so counter the perception that only men study computer science.

Actions aimed at getting women to accept their offered study place

One of the most successful actions in this category was the 'Women's Day', which sought to inform and motivate women applicants and make them visible to each other. All those admitted were invited to visit NTNU and offered a comprehensive presentation of the department and the university. Women working in the software industry, intended as role models, gave talks. The participants could also attend popularised science lectures, participate in an Internet course, and the like. They were guided by women computer science students during the weekend, and all expenses were covered by NTNU. The organisers put a huge effort into the Women's Day. According to the Pro-rector, they deployed 'a whole lot of seductive tricks'. For example, they used a popular and charming men's student choir to entertain and guide the young women around in Trondheim after the formal part of the presentation was concluded. The women were then invited to the choir's own premises in the House of the Students' Association where they were served strawberries and champagne.

Approximately 80% of the women who attended the 1997 Women's Day, entered the programme; 51% of the women students said it had been an important source of information about computer engineering (Berg and Kvaløy 1998). According to the WCI, the number of women applicants who accepted their offer, increased from 30-35% to 80% after this activity was established. Thus, the Women's Day was regarded by the WCI as one of its most successful actions. This was also confirmed by many of the women students we interviewed, who emphasised that they had been encouraged to enter the programme because of the Women's Day. In the language of inclusion needs, then, this was another powerful effort to increase the motivation of women to enter computer science; it also reinforced what was arguably the main message of the advertising campaigns – that women are wanted and they are welcome. The Women's Day was more about demonstrating desirable aspects of being a computer science student in Trondheim than an effort to fill information deficits.

Actions aimed at retaining women students

Several actions sought to improve the socialisation of women into the computer science programme and so reduce the previously substantial drop-out rate. Some minor curricular reforms were undertaken. The module 'Know your field of study' became part of the mandatory course for all first year students, who had to interview computer professionals at work. The purpose was to increase students' understanding of their future work situation and to make them see the relevance of their education – a clear case of seeking to redress deficits in how computer science had been taught and at the same time seeking to motivate students about the programme.

This action was based on the assumption that women students more than men are concerned with the usefulness of what they learn. However, our interviewees had mixed views about the effort. Some thought it had been interesting, while some thought it had been frightening:

> What scared me a little ... there was this lady who was talking [about the company] and it seemed very result-oriented. They had a performance assessment once a year, when they told people if they should find someplace else to work. But I don't think I would have been so scared today, but I was sitting there and knew nothing. We got the wrong impression ... I think they focused on the wrong things really. [...] We knew nothing when we started the first year. So thinking about working with computer science was actually a bit frightening (Fourth year woman student).

Another action in this category sought to improve the quality of the computer science programme. New technical equipment was provided and project-oriented teaching was implemented in two of the introductory courses. In addition, a new computer lab, Cybele, was established as an all-women lab. Unlike most of the other actions, this caused considerable commotion and provoked many men students at IDI. As a result, the Cybele lab was situated in a basement, a bit away from the rest of the Department's premises, in order to be unattractive and prompt as little attention as possible. The lab had 19 computers and was run by five women teaching assistants. They organised courses in HTML, Unix, PhotoShop, hardware, and the like to which only women had access. In addition, two women tutors were hired to provide general assistance and advice four hours a week. Clearly, these measures built on two deficits models: they sought to redress the presumed deficits in women with respect to confidence capability with computers; and they sought to overcome any 'minority problems' by providing a space where women could work with other women.

In the event, none of our women informants used the Cybele lab much; when they did, it was mainly to check mail and use the scanner. Normally, they would go to the other computer labs because they often collaborated with men students. Their views on this initiative were mixed:

> Vivian: Have you made use of the lab?
> Charlotte: Mainly the first year really. I have used the scanner.
> Carina: I have not used it, because I think it is so terribly unfair. I think I have been there once or twice. But I think it is terribly unfair.
> Camilla: I think I used it a lot the first year. But I haven't used it that much later, because if you are in a group with boys, you can't use it if you are going to sit and work there. But it would have been easier to find a machine there.

Christel: In a way I don't think it is unfair either. It's women who have created it and got hold of sponsor money in order to provide a milieu and well--being, and in a way … what we really have done is just to occupy a room. Anybody could have done that and got the same thing if they had wanted to.

Carina: But the point is that these boys hadn't been allowed to do the same thing.

Celina: No, but then they haven't tried either.

Carina: No, they haven't tried, but we do know very well that if they had, it would have been reacted harshly against.

Christel: That's not our fault.

Carina: Yes, but the point is that even if they had done the same thing, and got money and all that, they would never have been allowed because it would have been deemed discriminating. I am certain of that. Do you really think that NTNU had accepted that there was a computer lab where only boys were allowed? I can never imagine that!

Perhaps the most effective set of measures introduced with the aim of improving the retention of women in the programme provided all sorts of mechanisms of support – that crucial inclusion need which so often underpins the strengthening of motivation and capability (cf. Chapter 3). Two teaching assistants were hired to support well-being and community building among the women students. They were responsible for organising social events, and for inviting women professionals from industry to give talks. Another role model initiative was to hire a woman lecturer to teach the main introductory course of the programme. Social events included offers like sponsored travel to, and a one-day ticket, for 50 of the women students at a downhill skiing resort near Trondheim. This was very popular. In return, the sponsoring company Telenor was allowed to present themselves to the women. Another such event was a partly sponsored 'culinary evening' at a restaurant, where women students were offered to participate in making a three-course exotic meal.

It seems the teaching assistants' activities did contribute significantly to creating and supporting community building among the women students, and thereby reduced the number of dropouts. These measures may be seen as addressing the minority problem deficit in the culture of computer science by fostering a culture of mutual support and solidarity. But it must also be appreciated that the increase in the proportion of women students as a result of the quota in itself facilitated a much needed sense of community amongst the women students. Many made positive comments about their learning environment, which they saw as having improved mainly because women were no longer a small minority.

Other retention efforts included the introduction of an advisory system to help new students and introduce them to the programme by linking up with more experienced students. Another was a change in the way team-based exercises were organised. Usually, students worked in groups of eight, sometimes divided into subgroups of four. These groups remained the same throughout the year. As a result of the WCI, the groups were composed with emphasis on gender: if there were women in the group, half of the group should be women and all subgroups should always have more than one women student. This reform was generally well received by students, who appreciated being in mixed teams of men and women. And as part of its evaluation, the WCI manager conducted regular interviews or informal talks with most women students in order to learn more about them and how they experienced being in the programme.

Lessons to be learnt: Why was the WCI a success?

As we have seen, WCI comprised an impressively wide range of actions and activities. This chapter has analysed the most important ones – the outreach to girls in upper secondary school, the quota, the advertising campaigns, the Women's Day, some curricular reform and community building activities. It is impossible to identify any one of these instruments as more important than the rest. Indeed, one of the main lessons from the WCI is that it takes a broad and long-term effort to get more women into computer science.

Referring back to the four inclusion needs identified in Chapter 3, we argue that the main activities of the WCI met these needs in the following way:

- *Access*: the quota, the Cybele lab
- *Motivation*: various and complex. The Women's Day was perhaps most constructive because it supplied information that there would be more women students and that it could be interesting and pleasurable to study computer science.
- *Capability*: this was addressed by the advertising campaigns, but with limited success, and also by a variety of efforts to supply role models.
- *Support*: there were special teaching assistants for community building, including the facilitation of mutual support. Teamwork exercises were organised to make groups of half and half men and women.

Such a comprehensive set of measures may be necessary to meet the challenges to get more women into computer science effectively, and it seems that the different instruments reinforced each other.

Two other general factors contributed to the success of WCI. First, the initiative was well organised and well resourced, with explicit support from the top of the institution. Sadly, these ingredients are frequently lacking in diversity and equality efforts, with the result that policy commitments to gender change then have little impact in practice on the ground (Lee et al. 2010). The presence of these ingredients at NTNU meant that the WCI manager had the authority and funding to initiate a very large and comprehensive programme of measures, and undertake the associated activities needed (e.g., evaluations) to ensure that they are implemented effectively (cf. Lee et al. 2010). It also inevitably aided the visibility of the initiative.

The second general factor contributing to the success of WCI was the wide stakeholder backing which sustained support across the organisation. This was an important prerequisite of the WCI, since it involved an interaction of top-down and bottom-up efforts (see below). Significantly, the different stakeholders had different interests in the initiative: some wanted to increase the total number of computer science students while others wanted to increase the number of women students. This proved beneficial because it was easier to introduce potentially controversial inclusion measures for women – most notably the quota – under circumstances where a marked increase in the number of women could happen without making it more difficult for men students to be admitted. What emerged was a win-win situation which served to entrench the WCI, not just in the NTNU leadership but also in the department and among its students.

There is no reason to presume that this kind of win-win situation could not be replicated elsewhere, given that most countries are keen to increase the total number of graduates in computer science and engineering, due to the global growth of the industry. We have seen that it is possible to piggy-back efforts to include more women students on a wider strategy of growing the discipline, not the least since many high placed actors consider women to be an important reserve of talent that one should tap into. The upshot of this is that effective inclusion strategies for getting more women into computer science education require the political savoir-faire to build alliances. The key stakeholders at NTNU had differing views about what was the underlying problem causing the under-representation of women, yet they managed to meld their ideas into a fairly coherent and effective initiative. This pragmatic approach may have been a significant factor for achieving the win-win situation that this initiative benefited so much from. Thus, flexibility and pragmatism seems to be crucial prerequisites for a successful initiative, reflecting perhaps exactly the complex and multi-dimensional nature of the 'woman problem in computer science'.

Whilst the diversity of inclusion instruments employed was an important aspect of the WCI, some instruments should be stressed. Above all, it is clear that the quota for women was critical. This was an instrument that affected several dimensions of the inclusion problem in one move. First, it directly increased the number of women students admitted. Second, because it was such a radical move, it signalled that NTNU was very serious in its invitation to women. Third, the influx of women students made the computer science programme look different. As nearly all our student interviewees emphasised, the very fact of having many women in the programme created a more congenial student culture.

It has been argued that increasing the proportion of women in places where they have been a minority can produce a threshold effect (see, e.g., Etzkowitz et al. 2000, Oliver and Marwell 2001). The argument of this 'critical mass' theory is that once a particular threshold level has been reached, a dynamic process will start that inevitably leads to further increasing numbers. The theory has been criticised for the assumption that a critical mass necessarily prompts an accelerating enrolment of actors (Studler and McAllister 2002). Henry Etzkowitz and colleagues (2000) in their study of women in academic departments and research units discovered that critical mass did not work as a means to overcome isolation for women or as a catalyst for further enrolment of women. However, they suggest that critical mass did work in the sense that a relatively high number of women faculty meant greater support for women students in scientific departments (Etzkowitz, et al. 2000: 111).

These findings resonate with those reported in this chapter. On the one hand, it is clear from the WCI that a continuous effort is needed to maintain a high level of women applicants. On the other, it is also evident that increased numbers are vital – not just as an outcome but also as an instrument. It may look like a truism, but it is tempting to argue that, in the final instance, the most successful and critical inclusion effort was in fact to raise the number of women.

The contribution of higher numbers of women as an instrument of inclusion is evident in relation to the efforts to facilitate community building and create a more peer-supportive environment among the women students. The presence of large numbers of women in IDI made this far easier than it would have been previously, as it did in the study of Etzkowitz et al. No doubt, the activities to establish a community for women students at NTNU made a significant contribution to the positive outcomes of the WCI. Similarly, in her study of US computer science departments, Cohoon (2006) shows that peer-support improves women's retention in computer science. For much the same reasons, the importance of community building and peer-supportive spaces is also evident in our research into networks for women in ICT, described in the next chapter. Indeed, the value of 'women-friendly spaces' is a recurring theme in quite a diverse range of the digital inclusion strategies we have investigated, as we shall see in Chapter 9.

The networks for women in ICT may be characterised as bottom-up strategies, in contrast to the explicit top-down strategy employed by the WCI. These networks were created and run by women who largely worked in fields dominated by men. In many ways, they probably functioned as substitutes for the community and solidarity these women often lack in their workplaces. In a sense, the WCI achieved similar effects. By helping create a community for women, and through the substantial efforts to improve the learning environment, the WCI clearly had features of a bottom-up strategy, even if it was initiated from 'above'. In effect, the WCI top-down strategies were appropriated by its users, and the teaching assistants were appointed in ways that facilitated the women's efforts to perform student-to-student support. This may be seen as an important reason for its success in retaining women students as well.

Making positive circles

The analysis of the advertising campaigns raises the question of whether it is possible to change gender images through advertising, and if so, what are the critical features? Stereotypes are generally known as having a limiting and self-fulfilling effect on people (Correl 2004, Ridgeway and Correl 2004). It has been commonplace to describe computer science as 'masculine' and to offer the apparent symbolic unity between computers and masculinity as a main reason why few women enter the field (e.g., Wajcman 2004). The findings discussed in this chapter suggest that gender symbolisms are much more dynamic and fluid than this, since in practice much of the gender marking of a field is related in a quite straightforward way to the relative number of men and women (see Sørensen and Berg 1987; Faulkner 2001). The point may appear banal but its significance is profound: when men constitute a large majority in a field, that field gets coded as 'masculine'; when women dominate the resulting coding is 'feminine' (see e.g., Lagesen 2008). By the same logic, more equal numbers of women and men makes the gender coding more open; the field is likely to seem 'transgendered' (see Lagesen 2007, Blum and Frieze 2005).

This suggestion should not be read as a liberalist and reductionist view, that gendering is only a matter of numbers. Nevertheless, it is important to be aware that there is strength in numbers, that the relative numerical presence of men and women is very important to the way computer science is perceived. Moreover, one lesson from the WCI is that so-called image problem of computer science cannot be changed head on, for example, through creative advertising. The replacement of one particular stereotypical co-production of gender and technology with another one, which to some extent happened through the WCI, is not very effective. What is needed are measures that in a more direct way produce a gender balance, like quotas, or also by actions that directly change real appearances, like the provision of more role models. The actual effect of the NTNU advertisements was as a visibility strategy, showing that there really were many women in computer science and that they succeeded in the programme. Such issues will be further discussed in Chapter 10.

But how do we get women to apply besides by demonstrating that there are women in computer science, thus showing that this is a place for women? The findings from our study are quite clear. The women students said they were primarily attracted to the programme by its good career prospects and its high status. Chapter 5 analyses a different approach to inclusion, focusing on play, fun and glam rather than instrumental reasons. An interesting example with particular relevance to this chapter is the UK IT Beat project [Pitt 2003 a, 2004 a]. This was, as described in more detail in Chapter 5, a project to make young girls more interested in and less put off by IT, by glamorising it and associating it to men pop stars. This strategy was based on getting across the message by using fun and glamour as the main attraction.

While such approaches may be effective when addressing young women in general, our findings strongly indicate the importance of promoting instrumental arguments such as career prospects and job opportunities. Probably, it would also be a good idea to foreground the 'intellectual excitement of computer science' (see Roberts et al. 2002). At least the women students we interviewed were not particularly concerned about working with people, which was a basic idea in the WCI advertising campaigns, developed from a stereotype of what motivates women in general. Our analysis strongly suggests that recruitment campaigns should not target women in general but rather the women potentially interested in science and engineering and/or in careers in such fields.

We have also noted that many of the instruments were informed by all three types of deficit model outlined in the introduction; deficits in women, deficits in computer science, and deficits in the image and culture of computer science. Clearly, these models were important in shaping various inclusion instruments, but as we have seen, often with problematic results. The most conspicuous example was the advertising campaigns. Surely there are deficits that need to be corrected, but such models should be used with care as the basis of shaping inclusion instruments. As we learnt from Margolis and Fisher (2002), when inclusion instruments are made from observations of exclusion deficits, this should be done using an inclusion perspective. What does that mean?

The main bulk of the literature reviewed in the introduction of this chapter departs from observations that women seem excluded from higher education in computer science and related work. The questions asked circle the causes of this exclusion, viewing them as deficits related to the field of study, educational practices, etc. We claim in this chapter that such deficits models may be useful to identify problems but less so in shaping the remedies. The best example where the WCI made a wrong step, was in the advertising campaigns that conversed with deficits and stereotypes. The Women's Day on the other hand is a good example of an instrument constructed from a belief that there could be an information deficit but shaped by efforts to make the study of computer science appear attractive to women.

However, the really important achievements of WCI resulted from the making of positive circles of inclusion. By this we mean a process where various inclusion efforts link into each other to enhance the overall inclusion effect. Obviously, the quota for women played a principal role to get the process going. According to the women student interviewees, the quota was important mainly because they saw it as emphasising that the invitation to women students was seriously meant. The Women's Day allowed the prospective students to get first-hand knowledge about the place of study as well as to see that there actually would be many other women. The socialisation efforts to retain the women, like the social activities, arguably represented yet another reinforcing turn of the positive inclusion circle.

Such positive circles may not be easy to establish or to uphold. However, the concept is important because it serves as a reminder to be concerned with the way various acts of inclusion preferably should work in concert. This means that one has to consider how they relate to each other – or not. The making of positive circles represents an overarching ideal for the development of inclusion strategies.

This is also relevant to the way that the understanding of the gender-technology relationship among the actors involved in an inclusion project, influences the shaping of inclusion strategies employed by the project. Much of the literature on the gender-technology topic argues that most versions of this relationship works to exclude women (see, e.g., Wajcman 2004, 2009). Such claims invite efforts to criticise, correct and remedy exclusion forces, as the WCI did in their advertising campaigns by ridiculing hackers and promote an image of computer science made from a stereotype of women. Arguably, such efforts will fail because they do not help to create positive circles of inclusion. In the case of the WCI, it was largely ignored and deemed as caricature by most of the women students we interviewed. Only a small minority of women (or men) will ever consider studying computer science. To work from a general theory of exclusion to motivate this highly select group should be seen as yet another example how thinking in terms of one size actually fits nobody.

8

Empowering women in ICT through professional networks

Lisa Lee

Chapter 7 was concerned with strategies to increase the recruitment of women to computer science education and thence, hopefully, into specialist ICT work. This chapter is concerned with strategies to improve the retention of those women who do go into ICT work. Much of the literature on this question highlights the existence of a 'leaky pipeline', whereby proportionately more women than men leave ICT specialist work, at all stages of the normal career cycle. In particular, it seems, there are sizeable losses of experienced women ICT workers, mid-career, in many EU states. Retention is thus a serious problem, especially with respect to the more highly skilled and influential jobs in the sector. Whilst the proportion of women working in knowledge industries is relatively healthy, the proportion of women working in core ICT roles – in the design, reproduction and implementation of ICT products and services – is small. This situation is bad for women, who are not getting their 'fair share' of high status and high paid jobs (Koput and Gutek 2010). It is also bad for the sector, which is not getting the benefit of women's skills and, by failing to sustain a diverse workforce, may fail to produce the innovations needed to sustain growth in a diverse marketplace.

It is frequently argued that women are lost to ICT work because they experience ICT workplaces as a 'chilly culture for women' (e.g., Stanworth 2000, Valenduc et al. 2004, MacKeogh and Preston 2005). Carol MacKeogh explored these issues in a survey of 61 women ICT workers, most of them working in middle or higher level management roles in ICT companies in Ireland [MacKeogh 2003d]. When asked what they experienced as the main barriers to promotion in the ICT sector, her respondents pointed to two sets of factors: the lack of family-friendly work practices, and various disadvantages for women which occur when men are in a large majority in ICT workplaces. These findings are in line with the wider literature on the 'chilly culture' for women in engineering more generally. They would also seem to explain two of the common reasons cited for the loss of women mid-career (Lee et al. 2007) – that having children is a major impediment to women's careers in engineering fields, and that even those who do not have children often find themselves overtaken by men peers in terms of career progression.

The ICT sector is seen as less family-friendly than others because of an expectation of long hours, high commitment to the job and the need to continuously upskill in a fast-moving field. Some recent studies suggest that the long hours culture may not be as widespread as previously portrayed (Meiksins and Whalley 2002, McKinney et al. 2008). And certainly, the industry is seeking to improve work-life balance by promoting more flexible work practices (e.g., Department of Trade and Industry 2004), partly in response to employee pressure from men as well as women. However, according to HR managers surveyed by MacKeogh in ICT companies, the lingering perception that most high-level jobs require long working hours mitigates against the uptake of flexible working options, even when these are available [MacKeogh 2003f]. Progress in this area is slow, and as the responsibility for childcare and other dependants falls disproportionately on women, this remains a major factor impeding the retention and progression of many women in technological occupations (Lee et al. 2007, Perrons 2003, Bagilhole 2006). Having family responsibilities seems not to create the same kind of work-family conflict for men as it does for women in the IT industry (Koput and Gutek 2010:96).

Nevertheless, it is the disadvantages to women flowing from the fact that men are the large majority in most ICT workplaces which are the concern of this chapter. Following the terminology introduced in Chapter 7, we refer to these issues collectively as 'minority problems'. In brief, these relate to three main issues: (i) that the workplace cultures which shape every day interactions on the job are more comfortable for (more) men than women; (ii) that the social networks through which people learn the informal rules about how to get on tend to be 'boys networks'; and (iii) that women tend to be invisible and/or stereotyped professionally. Taken together these disadvantages can have a huge bearing on how much women feel and are perceived to 'belong' in technological work, and on how well their careers progress – both of which can have a bearing on whether or not they stay in ICT work.

This chapter investigates the potential of professional networks to enable women in ICT to overcome these minority problems, on the grounds that they have been found elsewhere to be effective forms of career support (van Emmerick 2004; McCarthy 2004a). We studied the nature and experience of three professional networks for women working in different areas of ICT: namely, WITI studied by Carol Mackeogh [2003d] in Ireland and myself in the UK [Pitt 2004b]; WOW in the Netherlands studied by Irma van Slooten and Els Rommes [2003, also Rommes and van Slooten 2004]; and Untold in the UK studied by myself [Pitt 2003a; Pitt 2004a]. The studies of these networks drew primarily on interviews with those responsible for the design, implementation and running of the initiatives, as well as interviews with a number of women members or participants and, in the WITI Ireland study, a survey of members. These studies asked: Why do women participate in these networks? What do they get from them? To what extent and in what ways can professional networks serve as a driver to greater inclusion of women in ICT?

A key feature to emerge from all three networks is that they *empower* women. Drawing on the work of Oxaal and Baden (1997) on empowerment, and McCarthy (2004a, 2004b) on women's networks, empowerment provided through women's networks is defined in this chapter as a grassroots process which enables women to challenge or overcome, individually and collectively, some of the minority problems which disadvantage them in ICT workplaces. We argue that the networks act as both "networks *of* empowerment", through their various activities and the messages they promote, and "networks *as* empowerment", through the nature of their organisation. Importantly, this analysis reveals some key elements that are needed to improve women's participation and advancement in ICT work. First we introduce the networks studied, and share some of our informants' experiences of being in a minority in ICT occupations.

The networks

Women in Technology International (WITI) is an international organisation, established in the US in 1989 with the aim of supporting women working in technology. The group's activities are now described as offering support, inspiration, education, conferences, publications and a wide range of resources for all women who consider technology to be central to their work. The WITI Thames Valley chapter (UK), set up in January 2001, was the first in Europe. It offers web-based resources and runs meetings on a regular (bi-monthly) basis, hosted by leading technology companies where its members work. These meetings may involve specific training sessions or talks by speakers. The WITI chapter in Ireland was established in 2002. The activities and aims are similar across all branches, although the frequency of meetings or topics of talks are decided locally.

The Dutch group 'Women on the Web' (WOW) was originally set up in 1996 as a branch of the US Webgrrls group in the Netherlands and renamed itself WOW in 2001. Its aim was to create an Internet-based network for women. The most important objective of the WOW founders was to make new media, such as computers and the Internet, more accessible to women. It seeks to achieve this by providing mailing lists, courses, and information on ICTs via its website. Its mailing lists cover computers and computer-related subjects, as well as catering for specific groups, such as enterprising women and women journalists. The group has received a great deal of interest, with around 4500 women subscribers in 2003. Although initially made up mainly of IT professionals and journalists, the WOW network now includes women from a variety of backgrounds. WOW emphasises that participants do not have to be IT specialists.

Untold emerged in reaction to the under-representation and invisibility of women in the field of digital design. It started as an exhibition at the Institute of Contemporary Arts (ICA) in London (UK) in the summer of 2002. The exhibition itself was initiated by the ICA, in response to a digital festival held there earlier that year, which revealed a huge disparity in the number of men and women participants showcasing their work. The Untold exhibition was subsequently complemented by a public debate reflecting on some key issues facing women designers. The high level of interest in both these events inspired the curator of the exhibition to set up a website as a platform for women digital designers to showcase their work. This idea has since been extended on: members now also share their experiences and socialise through the network.

Untold has thereby become a community of (predominantly women) digital designers, who seek to foreground their work and further their careers.

Being a minority in ICT workplaces

The existence of groups such as WITI, Untold and WOW tell us a great deal – through their work and the stories told by its members – about how gender imbalances in ICT-related work are experienced by women in these occupations. Asking women members of these networks why they participate in them inevitably produced testimonies about exclusion as well as inclusion, since it is precisely the minority problems identified in previous research on women's employment in ICT and engineering which motivated them to participate.

It is frequently claimed that the culture in many scientific and technological occupations is not welcoming to women (e.g., Greenfield 2002: 31). In her ethnographies of engineering workplaces, Faulkner (2009a) has shown how the largest cultural group tends to shape everyday practices at work – the subtle markers of belonging or otherwise performed through ways of interacting, topics of conversation, humour and social networks. Particular masculinities often set the norm for men and women alike, such that those who do not conform feel uncomfortable or marginalised in the culture; they don't quite 'belong'. Untold members described the existence and impact of a 'lads' culture [Pitt 2003b] in the new media industries. One woman designer told of men in her workplace emailing each other pornographic and offensive pictures, but not sending them to her. Although she did not have any particular interest in the photos, she was de facto excluded from the social life of the office because she was a woman.

The fact that there is a gender imbalance in terms of numbers, and that most of the men are young, means that this 'lads' culture is able to thrive. According to the founder of Untold, the problem is compounded by the managerial inexperience of many young men setting up small new media companies, especially as concerns gender policies:

> A lot of the people [men] setting up these companies came straight out of college and hadn't worked for very long, 'cause it's such a new industry and so they're not very well versed on sexual politics and what was acceptable in the workplace, [...] if you come from college I think you may be a bit naïve, and you don't realise that it's not OK to watch porn or have nude calendars in the office [Pitt 2003a: 246].

These young men tend also to recruit other young men like themselves, thus helping sustain this culture.

Several informants described how they felt excluded from workplace networks dominated by men – at times by default, because women have other out-of-work commitments, and at others deliberately, because men choose not to include them. As one WITI member explained, networking after work in the pub or on the golf course is not always an option for women, who "anyway do not get invited" [Pitt 2004b]. Culturally dominant groups of men not only shape who gets included and excluded in everyday interactions: they can also wield considerable organisational power. Untold members felt very strongly that women in their line of work often lose out on job offers, not because of a lack of skills, qualifications or talent, but because they are not in the same networks as the men and so do not hear about opportunities. As Untold's founder explained: "a lot of […] the recruitment policy is often word of mouth, and if you're not tapped into that network, you miss out on the job" [Pitt 2003a: 245]. Since jobs are rarely advertised in the flat hierarchies of new media, it is not 'what you know' but 'who you know' that matters (Gamba and Kleiner 2001).

Women's lack of access to such informal 'boys networks' is often cited as a key justification for having networks for women working in technological occupations (McCarthy 2004a). It is not just that men find out about jobs through their informal networks; they also learn how to operate effectively under the tutelage of more experienced mentors and role models in their networks. This may be particularly critical in a sector where, as the HR staff in Ireland reported [MacKeogh 2003f], pay and conditions are often negotiated individually, and where there are few formal or transparent procedures to ensure that promotion is based on merit (c.f. Robinson and McIlwee 1991).

There is some consensus in the literature that women in technological workplaces face what Faulkner (2009b) calls an in/visibility paradox: they struggle to be visible – or taken seriously – as technically competent professionals, but are 'hyper' visible as women (Carter and Kirkup 1990, McIlwee and Robinson 1992, Tonso 2007). This appears to be a common experience for women working in a range of ICT organisations and job functions, even in newer technology sectors (Gamba and Kleiner 2001, Perrons 2003), despite their 'egalitarian image' (Gill 2002, Pitt 2003). The invisibility means that women have to work harder than men to prove their technical ability; and they have to do this throughout their career, potentially every time they work with a new associate (Faulkner 2007, 2009a, b).

The impact of such processes on women's confidence is significant, especially in workplaces with only one or two women ICT workers. WITI's aim to educate and inspire women to reach their full potential [MacKeogh 2003d, Pitt 2004b] demonstrates that self-belief is a recurring problem for many women in ICT work. Often this lack of belief in women does not come so much from women themselves as from men colleagues and managers, even from partners and friends [van Slooten and Rommes 2003], who voice doubts that women are able or willing to take on senior and leadership roles. Similarly, in the Irish WITI survey, many women ICT workers reported that their men colleagues often see women as simply 'not up to the job' [MacKeogh 2004b].

Such gender stereotyping appears to be deeply engrained in many workplace cultures, even ones dominated by women. Simpson (2004) found that men working in occupations dominated numerically by women (such as nursing, air stewarding or primary school teaching) are more likely to progress to the higher grades, in part because they are supported to do this by more senior colleagues, but also because there are greater expectations upon them to do so. Similarly, one woman senior executive and WITI member described how gendered expectations play out in some ICT workplaces:

> Early in [women's] careers, people – men – are deciding for them that they will not want the overseas assignments, the riskier roles, the longer hours, to return to full time from their third child, etc. The culture in which they work is upheld, subconsciously in most cases, by the leadership which is 90% or more men.

The obduracy of gender norms and stereotypes heightens the visibility of (some) women ICT workers as mothers and carers in ways that contribute to the discriminatory impact of family commitments on individual career trajectories (Duerden Comeau 2004, Lee et al. 2007).

Networks of Empowerment

The very existence and growth of WITI, Untold and WOW clearly demonstrate that women do want to work in ICT, but that they also want to make it a more welcoming place for women, and to exert more control over their careers. The networks work to empower their members, and improve women's inclusion in ICT workplaces more generally, through three types of activities. First, they provide a women-friendly community to which the women belong unequivocally, and through which they can gain solidarity and support from others like themselves. Second, they seek to improve women's career progression and self-belief by supporting professional development in various ways. Third, the networks seek to give women in ICT greater visibility and recognition.

Community

As noted earlier, women are often heavily outnumbered by men in ICT workplaces, especially in the more technical and senior roles, and the culture is often not welcoming to women. Because the networks bring together large numbers of women, they are a helpful way to address women's isolation from other women at work. By interacting with others like themselves, the women come to realise they are not alone. They find strength in numbers (cf. Lagesen 2007), which help them deal with feeling marginal and fighting solitary battles in their workplaces. As one WITI member commented:

> It was so refreshing to me to meet other women who were working in [other] companies, who were not trainers or admin, because I hadn't met any. [laughs] ... through the network you realise that some of the challenges and emotions and impressions that you have aren't unique to you. [Pitt 2004b: 86]

These are women-friendly spaces. They provide a community that women ICT workers can and want to belong to, and they provide women a new way to connect to others who share their interests and identities around ICT. As one WOW user explained:

> I was looking for other women who are also fascinated by computers. I wanted something else than those men all the time. I had gotten tired of men [Rommes and van Slooten 2004].

For some, the networks are an alternative to, or a response to, other men-dominated networks – be they official professional networks or informal ones – from which they may be excluded. One WITI member told us she joined WITI because men in her workplace were members of organisations such as the 'Institute of Directors' [Pitt 2004b: 84]. Another saw it as creating "an old girls' network"; so countering the 'old-boys' networks, whilst providing many of the benefits such groups could offer.

The way in which these communities are experienced may vary, but the sense of solidarity is a common thread. All the women interviewed talked about the mutually supportive, trusting and safe environment the networks provide. WITI and Untold participants commended the groups for making them feel welcome, and other members for providing various forms of help in their work or in furthering their careers. As well as providing mutual support, members exchange all sorts of information and find new contacts, thus extending their personal networks, often across companies. As this WITI member experienced:

> I really hadn't done anything to develop myself in a long time and I thought, again, this was a way I could just see – is there a way, from a professional point of view, I could develop my network a bit more, my credibility a bit more, stuff like that. It actually has proved to be a big networking mechanism. But to me, what it is interesting is to see other women who are working for tech companies and hearing some of the challenges that their companies are going through and stuff like that. So again, you know, it's not just what's happening to you or your company [Pitt 2004c].

So, these networks are valued because they create communities of like-minded people; they bring together women doing similar ICT work, often with similar histories, and facing similar conditions, albeit in different organisations. Little wonder, then, that these communities generate a culture in which members readily feel they belong – in contrast to the experience of many in ICT workplaces. The positive community dynamics of networking and of mutual support and solidarity are self-sustaining, precisely because the participants share common interests and because, as we see below, they are empowered through their interactions with one another to become agents of change in their own careers and beyond.

Professional Development

As noted earlier, one consequence of being on the outside of existing 'boys networks' is that women in ICT jobs often lack access to the informal guidance in 'how to get on' which occurs through mentors and role models. This can be especially critical to career progression given that many sections of the ICT sector provide little or no formal training in professional development (Valenduc et al. 2004). Not surprisingly in this context, many members of these networks are motivated by a desire to further their careers – as was this WITI member, for example:

> The reason why I joined, was to become I guess a stronger person in the corporate world itself. So it's a great building block to kind of learn skills and be able to reinforce your own identity in a company, and that's certainly why I

> joined it. It's a great chance to potentially network as well, I have to admit but it's very much more (pause). I mean how to get up that corporate ladder and get recognised and not be afraid to kind of sing your own song, as it were, and play your own trumpet – which has to be done.

As this quote indicates, much of what these women need in terms of professional development is help with confidence, to believe in themselves and see professional goals as attainable. This is not surprising in light of the struggle to gain recognition for their ability and to counter the stereotyping noted earlier.

The acquisition of technical skills or core knowledge in their field of work is rarely an issue for the highly qualified women members of all three networks, although the WOW group does run ICT courses (such as webpage building) – because it has a more diverse membership occupationally and because one of their aims is to interest more women in ICTs and help women pursue careers in ICT. Apart from this course, the emphasis of the networks is on acquiring the 'soft skills', which all employees need in order to survive and progress at work, and which should further women's career advancement in ICT. For instance, WITI runs training sessions on such topics as goal setting, career strategies and body language. One interviewee explained that she chose to attend the WITI training sessions because other professional organisations offering career-related training and development do not address these types of skills: "You may have learnt something at WITI that you can bring into a job interview or bring into your next job" [Pitt 2004b]. Many of the skills learnt are not quantifiable; ordinarily, they are taken for granted or learnt informally in work or professional settings from peers or mentors. Yet, because women are relatively invisible and isolated in ICT workplaces, they might not get the same opportunities as men co-workers to learn from senior colleagues, nor get the same support to develop the confidence to promote themselves effectively.

Training is only one of the activities through which women members are helped to further their careers in ICT. In all three networks, formal and informal mentoring programmes and role models enable women at various levels to hear about and learn from the experiences of more senior women in the field. This is a key feature of all such networks (McCarthy 2004a). In addition, WOW also provides websites with on-line learning resources, and information about successful women in ICT [van Slooten and Rommes 2003]. And Untold seeks to inspire its members through public debates and special features, including examples of work by talented women digital designers [Pitt 2003a].

Much of what women gain from these activities – especially the networking and role modelling – is inspiration and encouragement. Arguably, these are vital elements empowering women to become proactive in their careers, especially when these women may have low self-esteem and confidence professionally. Meeting other women, much like themselves, who have progressed in their careers and learning from their examples, can provide the spur to (re-)build the confidence needed to move out of a career dead-end and pursue a path that will fulfill their own personal aspirations. As these comments by WITI members illustrate:

> I think actually I've been quite surprised about some of the people here, who've been on the Board, how confident they are in their jobs, how they've moved up their career ladder, made career choices to move sideways or completely out of what they were doing. That's, I find that quite interesting and encouraging that people can do that and survive in the job market [Pitt 2004c].

> After that first meeting, discussion, I got a mentor through WITI who has been incredibly valuable for helping me personally to (pause) I'm developing the business environment and also to bounce ideas off other people in the development side as well. So that was really useful. As I said, the networking side has been really valuable, so I now have some really good contacts, both the speakers that have been and presented at these sessions, and also other members of WITI [Pitt 2004c].

It is no accident that women are learning and gaining inspiration and confidence in an environment which participants experience as safe and supportive, and where women feel equal among peers. Again we see the importance of community building and dynamics to the inclusion processes operating here.

Visibility

As well as offering women in ICT a sense of belonging and support in their careers, these networks also raise the visibility of women in ICT in new ways. The networks seek to make women in ICT visible to a number of audiences: to one another, to other women and men working in ICT, and to the wider society.

The visibility efforts are strongly anchored in the community building dynamics of the networks and mediated through a number of activities – including face-to-face meetings of members and non-members, the websites, and the network of contacts they create across organisations.

Many of the women participating in the networks would never have met one another otherwise (in so much as they either work in different companies or are not members of other men-dominated networks). The networks not only put women ICT professionals and users in touch with one another, they also indirectly draw attention to the existence of these women among their men ICT colleagues and users. By providing a platform for members to be heard and seen, and by bringing to the fore women role models and women in senior ICT positions, they help to refute stereotypical assumptions about the interests or abilities of women and go some way to tackling the in/visibility paradox discussed earlier.

The networks' websites have been a key instrument facilitating the visibility of the networks, and of women in ICT. These serve as a means of introducing new women to the groups, as a source of information for its members, and as a public space for discussion and exposure. The importance of the latter should not be underestimated. As we argued in Chapter 4, WOW has clearly given visibility to women Internet users and made their presence known. By bringing thousands of women on-line, they have challenged the image of the Internet as a 'toy for the boys'. WOW also engages in visibility activities directly targeted at the wider society: they have maintained a high profile in the press and on the Internet, with articles about women in ICT; and they have organised campaigns, for instance, against the producers of a sexist computer commercial. The audience Untold targets is the design and ICT industry. It uses its web presence to showcase the work of talented women digital designers to the rest of the industry [Pitt 2003a]. This was a founding motivation of the network: it provides women members with a rare opportunity to be heard and seen in their professional setting. Much as occurred in the Women and Computing Initiative described in Chapter 7, their efforts have made visible the significant numbers of women who are in ICT, so serving to 'normalise' the very idea of women ICT professionals more widely – as well as promoting individual women's work. At the same time, the publicity surrounding the initiatives and campaigns themselves brings visibility to the issues surrounding women's under-representation in ICT and related fields.

Untold in particular has a campaigning agenda around its visibility efforts. One member emphasised the group's efforts to engage all levels of industry, including designers and company directors, to discuss issues facing women working in the field:

> It's important to be changing the situation – rather than just focusing on what the situation is and those experiences and statistics about the way it is

> – focusing on ways that you can change it [...] Untold is making a difference and changing it [Pitt 2004c].

Engaging senior staff is indeed a critical step in changing working conditions and workplace cultures, given that many of the young men heading new media companies are, as noted earlier, seemingly unaware of the impact of a 'lads' culture on the few women employees. Untold and its members believe that raising awareness of the issues affecting women, and seeking solutions directly with employers, through dialogue, can lead to a more gender-aware working environment in ICT workplaces.

Finally, the women in these networks hope their efforts will have an impact on other women considering work in ICT. This WITI member described how she hoped the network would work in the future:

> [...] continue to be informative, creative, exciting in the ideas that it presents for women to get involved, in providing radical views on how we can better change the organisations that we are in to make it easier not just for ourselves, but for those women who have not yet graduated from university even [Pitt 2004b: 88].

Women participate in the networks, therefore, not just to get support in their work and careers, but also because they want to help other women working in ICT [see also Rommes and van Slooten 2004]. In both areas, the strength in numbers of these networks is again vital. It gives weight to the public presence of these networks, to their efforts to challenge the status quo, and as a voice for the concerns, ambitions and successes of women in ICT.

Networks as empowerment

We have seen that these professional networks operate as networks of empowerment because, by addressing minority problems experienced by women in ICT work, they empower members to become agents of change in their working lives and beyond. At the same time, the very form of these networks and the ways in which they operate makes them empowering for women; these are networks as empowerment too.

Bottom-up mechanism

The WITI, Untold and WOW networks – like women's networks in other occupations and industries – were set up and are run by women working in the field, for other women in the field. It is in the very nature of networks that they bring together and build communities of like-minded people, with shared identities and interests, thereby providing a space in which participants can feel welcome and belong. The safety of these communities allows their members to confront and share their experiences and ideas. The discovery of common ground serves to validate and acknowledge feelings that might otherwise remain private. According to Castells, networks by their very structure facilitate the convergence of the "affirmation of an individual-centred culture and the need and desire for sharing and co-experiencing" (Castells 2004b: 223). One WITI member's comments reflect this:

> WITI helped me to pull together my thought processes. I had some belief systems already that existed. But talking to other women and hearing the presenters, has kind of helped me to formulate more clear views on this topic [Pitt 2004b: 87].

The collective knowledge and energy generated in this way are what empowers and enables participants to address their individual needs, to provide support and solidarity to one another, and to further their shared aims. Sharing and learning are characteristic features of networks.

Because they are grassroots or bottom-up types of organisation, the networks are very close culturally to their participants, highly relevant and, as one interviewee put it, "very democratic" [Pitt 2004b]. Participants contribute to shaping the objectives and activities of the group, as well as helping implement them through their very involvement. In their analysis of gender and empowerment, Zoë Oxaal and Sally Baden (1997) view women's empowerment as necessarily a bottom-up process, rather than one that can be delivered top-down. Drawing on the work of Rowlands (1995), they explain that this approach is necessary since "empowerment is not only about opening up access to decision making, but also must include processes that lead people to perceive themselves as able and entitled to occupy that decision making space" (Oxaal and Baden 1997: 3). They argue that women must therefore empower themselves and work together to challenge established norms and gendered dynamics, rather than rely on external agencies (ibid: 10). In a similar vein, Helen McCarthy's study of other women's networks concludes that these are indeed a powerful way for women to generate their success through their own agency (McCarthy 2004b: 124).

This is not to say that organisations employing ICT professionals can do nothing to improve the retention and progression of women in ICT.

Quite the contrary: there is a raft of 'top down' measures which are known to help in both regards. However, a recent study of European engineering organisations found that the uptake and impact of such measures is often very limited: there is a widespread and serious gap between policies on paper and actual practices on the ground (Lee et al. 2010). In other words, the bottom-up interventions of these networks, and the emphasis on empowering individual women to become active agents of change is occurring in a context where employers have not done enough to make their working practices and cultures truly gender inclusive.

At the same time, there is considerable antipathy amongst both women and men in ICT, and in technological occupations more generally, to some kinds of gender equality policies – in particular, to career support provided solely 'for women' and to any form of 'positive discrimination'. Our own interviews and other research (Womeng 2005, Lee 2007) have revealed a fairly widespread perception that positive discrimination unhelpfully stereotypes women as needing help and incapable of success under their own steam, thus reinforcing notions of women being 'not up to the job'. For women who enter or are promoted through any kind of quota, this can impact negatively on their own sense of worth, as well as their career progress, as one WITI member found:

> ... there was like positive discrimination towards employing more women, to get the numbers, and I think I came across one company that I joined and I felt afterwards that it had been positive discrimination to give me the job where maybe I wasn't suited for that role. [...] Well I didn't stay too long in the job [Pitt 2004b: 80].

Nevertheless, Lagesen's [2004] study of the Women in Computing Initiative indicates that antipathy to quotas can be tackled if people are persuaded of the need for some sort of intervention and if the scale of the 'lowered threshold' is considered proportionate. Similarly, McIlwee and Robinson's (1992) research in engineering companies shows that affirmative action on the recruitment and advancement of women can be effective without undermining women's visibility or job satisfaction.

Transparent, flexible and open

The organisational form of networks, and the fact that this enables the sharing of experiences, resources, ideas and knowledge, makes them an extremely powerful means for achieving their aims (Castells 2004b: 224). In the case of women's professional networks, this power is exercised through transparent structures and formal status (McCarthy 2004a) – in contrast to the informal men-dominated or 'old boy' type networks described earlier, which are far from transparent or officially recognised.

This does not mean that women's professional networks "aspire to achieve the same hegemony for women that men have traditionally enjoyed in the workplace" (McCarthy 2004b: 122). Rather, their aim is to create a more gender-aware working environment, where issues affecting women, and men, are better recognised, and to share this understanding as widely as possible in order to guarantee the advancement of women's careers now and in the future. In her study of other such networks, McCarthy finds that:

> ... by actively creating opportunities for women to extend their personal and professional networks, women's networks alter the dynamics of the patterns of social ties established by men, and create alternative spaces in which male norms and behaviours can be challenged (McCarthy 2004b: 123).

The emphasis of these networks is therefore on notions of choice and opportunity; they are "fundamentally concerned with altering social relations rather than preserving them" (McCarthy 2004b: 123).

In line with this ethic, and the desire for transparency, the Untold and WITI networks have a clear policy not to exclude men from participating in the networks. WOW, by contrast, attempts to keep its on-line communities women-only, which they see as critical to providing a space where women can feel safe and confident [van Slooten and Rommes 2003]. For Untold, the principle of being open to men helps reinforce a message of inclusion [Pitt 2003a], even if very few men participated, and on the whole men did not feel targeted by the networks [Pitt 2004c]. In practice, this inclusive message is difficult to convey and the networks' aims are not always seen in this positive light, as one Untold member reports:

> People just don't get it. They think that it's about women designers trying to show off and say we're better than men (pause) and the point we're trying to get across is, it's nothing to do with men against women or anything [Pitt 2004c: 100].

Both Untold and WITI women members were concerned to further their careers without creating new boundaries of inclusion or exclusion and new barriers between men and women working in ICT.

There are further considerations here. One obvious one is that men are an important audience for these networks. In addition, keeping these networks open and transparent makes participation in the networks somewhat less problematic for many of the women, who appear to be wary of being branded feminists. It is not that these women see gender as irrelevant to their situation, as transpired in Gill's interviews with women in new media work (Gill 2002). Rather what seems to explain this avoidance of 'feminism' is that women working in men-dominated areas of technological work are highly aware of a sensitivity around placing too much emphasis on gender. They have learnt that by refuting or playing down the significance of gender, even describing the field of technology as 'gender neutral' (Jorgenson 2002), they are better able to fit in and maintain their professional identity (Jensen 2005, Carter and Kirkup 1990); playing up gender can be seen as a threat to the unity of the profession (Jorgenson 2002). There is perhaps too a certain stigma attached to earlier women's movements and feminist activism, which is out of tune with current workplace politics (McCarthy 2004a) and with public opinion more generally (*Guardian*, 2 July 2003). As one member of Untold explained: "A lot of women of our generation are nervous about appearing to be feminists" [Pitt 2004c: 102], where 'feminist' seems to imply engaging in a rhetoric of confrontation.

In any case, an emphasis on 'women's issues' is not always necessarily appropriate, since some of the career obstacles faced by women are not gender- or sex-specific. As one WITI member remarked:

> There's been some very, very specific learnings out of [the sessions], like how to dress to be successful in business, that type of thing, goal setting. And there was a very good session where we had some coaches, who came in, who talked about, for example, how to manage your career in a down market. Now, again, I don't necessarily think that was centred upon being a woman. I think that would have been valuable to any sex at that meeting but it happened to be at the WITI meeting [Pitt 2004b: 87].

This underlines a point made throughout this book, that many inclusion measure can benefit everyone – men as well as women – since they can improve working conditions or career prospects for all.

Finally, the activities organised by the networks tend to be relatively flexible in format and open to all. This means participation can be on an ad hoc or last minute basis – it is left to members to come when they can or when they feel the need. Participation in WOW is especially flexible, since the major part of their work is carried out through on-line mailing lists. No formal membership enrolment is necessary for either WITI or Untold, who charge instead a small fee for entrance to each event but offer free access to the websites. WOW subscribers do not have to pay, but if they do wish to become members, a small contribution is required.

Conclusions

The voices of WITI, WOW and Untold have provided vivid illustrations of how women can experience inclusion through networks specifically set up and run for and by women ICT professionals. These network are providing communities where women can build their networks of contacts, exchange information and find mutual support; they are supporting professional development by building up confidence and raising career ambitions; they are inspiring members through role models and mentors; and they are raising the visibility more widely of the women who do valuable jobs in ICT, so challenging stereotypes about women's place and abilities in this technological sector. In all these ways, the networks are empowering women members to become agents of change – in their own working lives and beyond, for women in ICT more generally.

We have seen that these networks help women obviate to some degree the various disadvantages associated with being in a small minority, as women, in the workplace – their lack of access to powerful networks, their marginality in the workplace culture, and their invisibility professionally. But the significance of women's networks goes well beyond providing individuals a form of support they are otherwise denied – powerful though this undoubtedly is. By giving voice to issues affecting women in many areas of technology work, by raising the profile of the number and talents of women ICT professionals and by campaigning in various ways to increase the number of women in these occupations, these networks are also engaging in a wider project of gender inclusion in the Information Society.

The inclusion processes we have witnessed here may be described in terms of inclusion needs of the women participating in these networks: They are highly motivated to get involved – not least, to improve their own situations at work. The networks provide them with the (soft) skills they often lack but need to further their careers. And the support they gain from one another is both emotional (inspiration and encouragement) as well as practical (contacts, advice, etc). But the critical dynamic operating in this inclusion mechanism is that it creates a community of individuals who (to varying degrees) share these inclusion needs, and therefore share a common interest in improving the situation for all women in ICT occupations. The more they gain from the network, the larger it becomes, and vice versa. As we have seen repeatedly, it is the 'strength in numbers' of these networks which makes them a powerful means of achieving their aims.

It is in the very nature of networks that they facilitate the sharing of experiences, resources, ideas and knowledge within communities. Other organisational aspects have helped: their transparency and openness is vital to the visibility efforts, as well as being consistent with a politics of inclusion. Their open and informal format offer a very flexible means of participating. But above all, it is the 'bottom-up' character of networks which makes them such effective instruments of empowerment and inclusion.

The experiences described in this chapter reveal some of the key elements needed to tackle the minority problems facing women working in ICT occupations and industries. There can be little doubt that these networks do improve the retention and advancement of women in the occupations designing and implementing ICTs. Nevertheless, there is still a long way to go before gender equality is achieved. Whilst the women who participate in these networks have indeed gained knowledge, inspiration and confidence to make a difference in their own work and careers, there are real limits on what women can do at an organisational level – for example, to make workplace cultures more inclusive – if they remain in a very small minority.

There is a crying need for more top-down intervention – in promoting 'culture change' (Liff and Cameron 1997). This could include measures such as those that actively encourage women into senior positions and reward publicly the contributions made by women employees, clear and transparent procedures for recruitment, diversity training as well as family-friendly policies, including tackling the long hours' culture endemic in ICT and improving flexible working arrangements and its availability, supporting equally career development opportunities for men and women through (re-)training and coaching, greater transparency and unbiased expectations as well as supporting the vital work of those groups, networks and individuals who are seeking to improve women's participation in ICT – *in addition to* supporting the bottom-up efforts of these networks and their members. There is indeed little hope that ICT workplaces will become more welcoming to women and supportive in their career development without real corporate support backed up by adequate resources from the organisations in which women work (Lee et al. 2007 and 2010).

9

A room of their own? Inclusion through women-centred ICT spaces

Knut H. Sørensen and Wendy Faulkner

Some of the digital inclusion strategies we studied were geared to 'everybody' while others were 'for women' only or chiefly. The latter included four rather different types of strategies: women-only ICT training, women's networks, websites and ICT products designed for girls or women. Only in the first of these is access (in any practical sense) restricted to women only, but in all cases the express intention is to reach women. Our analysis indicates that this represents a major class of digital inclusion strategies, with important commonalities in spite of the obvious diversity. In particular, they all involve the creation of what we would characterise as 'women-centred spaces'. This chapter seeks to compare and contrast the four types of 'for women' inclusion strategies we studied, in order to understand the potential of women-centred spaces to facilitate gender inclusion in the Information Society, and to reflect on any potential limitations.

All four inclusion strategies were designed with an explicit concern for 'women', and most sought to provide opportunities for women to encounter and explore ICT together with other women. In this sense, they all pursued a kind of category politics (see Chapter 1). Indeed, the starting point of our analysis is that 'for women' digital inclusion strategies represent an exercise of category politics, and that the main alternative to such strategies would be based on gender mainstreaming through gender sensitive inclusion initiatives aiming at 'everybody'. Drawing on observations from previous chapters, this chapter therefore reflects on the implications and relative merits of thinking from the standpoint of category politics and thinking from that of gender mainstreaming. Put more prosaically, we ask: Under what conditions are 'for women' strategies an effective means of increasing the women's inclusion in the Information Society? Or are 'for everybody' strategies generally preferable? In light of our general critique of gender essentialist thinking, we have a particular concern to explore whether the making of women-centred spaces can avoid some of the weaknesses of category politics, especially the tendency to reproduce gender stereotypes or essentialisms. Finally, is it possible to achieve some of the benefits of women-centred spaces within a framework of gender mainstreaming?

The body of this chapter addresses our four types of women-centred strategies, in the order in which they emerged historically. First, we review some relevant findings and arguments in favour of women-centred digital inclusion strategies.

Women-centred spaces and ICT?

The usual rationale for creating women-centred spaces lies in the belief that women may benefit from access to arenas where men are absent or constitute only a very small minority. This was a major tenet of early 'second wave' feminist activism in the 1970s, which placed considerable emphasis on autonomous women's organisation (e.g., within trade unions and students unions) and women-only consciousness raising groups. These spaces were seen as a necessary means to achieving real change in gender power relations, by enabling women to better understand the nature of their oppression, and to build the confidence and solidarity necessary to engage in collective activism to overcome it (Bondi and Davidson 2002). So, building women-centred spaces was seen as a necessarily transitional strategy in the broader project of gender equality.

Studies of specific kinds of women-centred spaces highlight a range of benefits. Kim (2002) compares women graduates from women-only and coeducational US colleges and finds that women-only colleges have a significantly greater impact on the development of intellectual self-confidence, but make no difference to the development of critical thinking, analytical and problem-solving skills. UK studies find that women-only trade union courses increase women participants' self-confidence and provide opportunities to identify participants' experiences of sexism and racism (Greene and Kirton 2002). Kirton and Healy (2004) conclude that, rather than leaving men's power intact, women-only training empowers women to become politically engaged to fight gender inequality:

> In short, the women-only educational context provides a safe site in which to develop knowledge, skills and confidence to tackle injustice and to use collective agency. It is also one in which individual women stewards and representatives find their gendered active/politicised identity reaffirmed and revitalised through the interaction with other women activists and members (Kirkton and Healy 2004: 320).

Concerns about empowerment and the need for 'safe sites' were also raised through early feminist research on computers and work, which saw computer as a potential risk for women's employment and quality of working life (Lie and Rasmussen 1983, Cockburn 1985, Probert and Wilson 1993, Webster 1995). The new technology was perceived as offering new opportunities for men but not women, by privileging skills usually attributed to men. For these reasons, feminists argued, women-only training was needed to provide an arena to challenge the gendering of the new technology and to overcome any gender-linked disadvantages. These arguments drew on widely reported evidence that many girls 'practise computer inexpertise' in mixed-sex schools (Volman 1997; Volman and Dam 1998) in case appearing too computer skilled and confident seems 'unfeminine' (Kaminer 1998). The hope, then, was that women-centred ICT spaces would not only provide safe spaces in which women would be empowered to benefit from the ICTs, but that they would also serve to neutralise or even reverse the predominantly masculine coding of ICT.

In Chapter 2, we argued that the symbolic coding of technology as either masculine or feminine rests in large measure on whether the technologies in question are predominantly used by women or men (see also Chapter 7). By this logic, we would expect that, at least within the confines of a given women-centred space, the gendering of ICT would change. Feminists are split on the question of how easily the gender coding of ICT can be changed. Judy Wajcman's (2004, 2009) overviews of feminist research on technology struggles to avoid giving the impression that, generally speaking, technology is masculine and thus works to exclude women. A more optimistic view has been offered by what gets called cyberfeminist approaches.

Cyberfeminism seeks to support and encourage women's exploration and domestication of computers. It is a strategy to use computers as political tools for feminist activism, while at the same time celebrating women's achievements with regard to ICT (Wilding 1998, Everett 2004). A central theme is the claim that, by its very nature, ICT creates a kind of tailwind for efforts to change the gender coding of ICT. Specifically, cyberfeminism foregrounds the potential of networked computers to provide new opportunities for communication, above all through the Internet. Communication is portrayed as a particularly 'feminine' skill, a 'feminine' alternative to skills and interests seen as 'masculine', such as control and information. For this reason, it is argued that the 'communication turn' in digital technology provides a gateway for women into computing.

Sadie Plant, perhaps the most vocal advocate of cyberfeminism, argues that technologies like the Internet, cyberspace and virtual realities have created an ideal 'feminine' medium, and that current developments are working to the long-term benefit of women (Plant 1997, 2000). This could be viewed as a strategy whereby ICTs are domesticated by women through labelling them 'feminine', or at least through making these technologies appear 'gender authentic' (cf. Faulkner 2009b) for women. In this way of thinking, creating women-centred ICT spaces is a strategy that would be expected to benefit from a positive tailwind. And certainly, the tailwind argument finds some support in research which highlights communication aspects of ICT as perhaps the most important quality of digital technologies facilitating the inclusion of girls and women as users (Håpnes and Rasmussen 2003), even encouraging them to become computer enthusiasts (Nordli 2003). As we saw in Chapter 5, the opportunity to use computers to communicate, to download music and do fun things on the Internet appears to have attracted more girls and women to ICTs, so opening up at least their symbolic coding; the technology is no longer so obviously 'masculine'.

This argument about the gender coding of ICTs resonates with ideas about doing 'feminine' designs – arguments made, for example, with regard to computer games (Cassell and Jenkins 1998b, Kafai et al. 2008) as well as pedagogical experiments (Brunner and Bennett 2002). In line with the framing of category politics, these positions are all based on an explicitly dichotomous understanding of gender. Brunner and Bennett's research about educational uses of ICT leads them to a clear-cut distinction between 'feminine' and 'masculine' relationships with the technology:

> In our research about the educational uses of technology [...] we have come to recognise two very distinct attitudes toward technology [...] One seems recognisably 'feminine' because it supports and reflects the values associated most strongly with women. It contrasts with a more traditionally 'masculine' perspective about technology [...] This feminine perspective regards technology almost as a fellow creature, needing care and feeding to stay in balance with the environment, both social and physical, that surrounds it. A feminine perspective wants technology to be small and versatile, to permit conversation and connection, the sharing of ideas and feeling and experiences. [...] The masculine perspective asks that technology grant the user transcendence over time, space and the limitations of the physical body. It sees technology as affording power, command, and control (Brunner and Bennett 2002: 71–72).

This argument could be interpreted as inviting a cyberfeminist strategy to get girls/women included into ICT by providing a 'feminised' space where girls/women may domesticate ICT without feeling that they enter into a 'non-feminine' or even 'masculine' domain. In this reading, the entry of 'feminine' women will change the gender coding of ICTs, where other cyberfeminist writing tends to imply a reverse causality: give these technologies a 'feminine' label and this will empower women to enter ICT.

The above discussion raises interesting empirical questions, which we explore in this chapter, concerning how effective women-centred ICT spaces may be, not only as a means of getting more women to become users and designers of the technology, but also as a means of challenging the masculine coding of ICTs. In each of the four types of women-centred strategies studied – training, networks, websites, and ICT products – we consider: Why were women targeted, and which women? How was the category 'women' understood, and what was achieved through such category thinking? How was gender constructed? What were the mechanisms and measures adopted? What was the overall impact of these strategies, both on women's ICT practices and on the gender coding of ICTs? And are women-centred spaces (still) a necessary step towards greater gender equality and inclusion?

Women-only ICT training: A stepping stone to inclusion

The 1980s saw the spread of women-only vocational courses in information technology (IT). In an unusual confluence of feminist and economic development agendas, public funding bodies across Western Europe were persuaded that women could be an answer to pressing skill shortages in the burgeoning IT industries, and that women-only training was a justifiable route to providing those skills. By 2000 the strategy had fallen from fashion. Still, we found two current cases to investigate: the Vrouwenvakscholen (VVS) in the Netherlands [Rommes 2003a, van Slooten and Rommes 2004] and the Edinburgh Women's Training Course (EWTC) in Scotland [Faulkner and Kleif 2003, Kleif and Faulkner 2004].

The VVS and the EWTC courses were the first of their kind in their respective countries, and share similar histories (for a fuller account, see Rommes et al. 2005). Both were founded in the mid-1980s, to enable disadvantaged women to enter 'non-traditional' (i.e., men-dominated) occupations in the IT industry. With increasing funding difficulties during the 1990s, the original ambitions of the courses were diluted somewhat, in terms of their technical content. At the time of our studies (2002 to 2003), VVS trainees chose from one of three options: web developer, web master or system/network administrator. EWTC trainees gained the European Computing Drivers Licence, which qualified them in operating most office applications of ICT. Nevertheless, these courses boasted considerable success in terms of both digital and social/economic inclusion. Both courses were designed and targeted not on women in general but on particular groups of women: unemployed women who would find it difficult to make the transition onto the labour market and/or are disadvantaged for various reasons (lone parenting, race/ethnicity, age, disability). Eighty per cent of the VVS graduates found a job within a year, albeit predominantly in low paid ICT jobs with limited options for promotion. Over half of the EWTC graduates went into a job or further education within a year.

Beyond these destination statistics, our interviews with former trainees revealed that these courses can change lives in quite profound ways. Many of the trainees were bringing up young children on their own, with very little money. Some had serious health issues and many were suffering from depression. Although the trainees were very diverse, what they all shared to some degree was a lack of confidence and self-esteem when they came on the course; most undervalued themselves and had very limited aspirations. This is common amongst long-term unemployed people, women returners (Rees 1992) and women in low paid marginal jobs (Brew and Garavan 1995, Simmons 1989). For all the graduates, the sense of achievement in completing the course was in itself hugely empowering. Caroline, a lone parent with a chronically sick child, told us

> I am not the same person! [...] Besides learning all the stuff on the computer, the biggest thing was that it just gave you so much confidence. [...] Nothing would faze you anymore, do you know what I mean? Nothing seemed daunting anymore.

And Begum, an Asian Scottish trainee who had become isolated after her children were born, learnt how to be with people again, to laugh and joke. She began to value herself again:

> It has increased my confidence level – that's the main thing. And then it's made me want to do things for myself: exercise, socialise. Before, my life was just a mess. It's picking up now.

Greater confidence and self-esteem can empower women towards a range of inclusion outcomes, social as well as economic. From this new base, women are more likely to be able to take steps to improve their circumstances. In these and similar courses (Stanley 2003, Cardiff Training Centre 2004), we see that improving self-esteem and confidence is a pivotal link between vocational ICT skill acquisition and wider inclusion outcomes for women who are unemployed and/or disadvantaged. How then is it done?

The training model of the VVS and EWTC worked because it offered a heterogeneous package of measures, which met the particular circumstances and inclusion needs (see Chapter 3) of the target group 'as whole people', in four key ways. First, the women were offered access to computers and to relevant training, and they were motivated to take up this opportunity: they saw ICT skills as relevant to them, something they needed both on the labour market and in everyday life. The courses had close connections with potential employees, which ensured that the skills they taught were up-to-date and in demand. In addition, trainees went on a work placement, which gave them a 'real life test' of their new skills, and helped them make the transition from the women-only course into a mixed-sex working environment.

Second, both courses included modules on 'complementary skills' alongside those on ICT. These elements were intended to provide necessary skills for job seeking and job keeping – such as verbal and written communication – but also to encourage personal development. As with other women-only courses, the latter included the sharing of personal experiences, the aim being, in the words of one trainer, to inspire trainees to "nurture their own spirit, to realise their full potential [and] take an equal place in society". It is clear that this very heterogeneous aspect of the training contributed hugely to empowering and enabling women trainees to change their lives.

Third, both courses accommodated the practical needs of their trainees. The training, and (at EWTC) childcare, are provided free of charge, and training takes place during school term times. For the two thirds of the trainees who are mothers, these are vital preconditions to their being able to participate. In addition, all of those we interviewed clearly gained emotional as well as practical support enabling them to complete the course. They felt they could talk to the trainers, even about personal problems, and valued the attention paid to their circumstances outside the course – all of which helped them cope with the transitions they were making.

Last but by no means least, several features of the organisation and delivery of the VVS and EWTC courses worked directly to build trainees' confidence and self-esteem. The marketing of the courses told the target women they are welcome here. The selection process did not 'set people up for failure', in that they selected women who, as well as being unemployed and/or disadvantaged, also had sufficient aptitude and motivation to get something from the training. The delivery of the training was modular to allow the trainees to progress at their own level and pace; formal training materials were supplemented by one-to-one and group support, delivered flexibly. All achievements were celebrated; trainees were constantly reassured and encouraged to take pride in their progress, however modest. The trainers sought to foster a safe and supportive environment, in which trainees supported and encouraged one another. The trainees acted as positive role models for one another: 'if she can do it, maybe I can too'. Women trainers also provided a strong demonstration effect.

A similarly heterogeneous package of measures has been found in other studies of women-only technology training. Moreover, many elements of the training model described here also worked well in four other mixed-sex community-based ICT courses studied [MacKeough 2003b, Hynes 2004, Rommes 2003a]. In particular, the celebration of achievement and the fostering of a non-competitive, supportive learning environment also appear to work for unemployed men in similar circumstances as the target women. This begs the question: to what extent and in what ways was the women-only aspect of these training courses instrumental in achieving their successful outcomes?

A review of literature on the topic (Brew and Garavan 1995) concludes that confidence building is easier to achieve for women in a women-only training setting than in a mixed-sex one. As one of the EWTC trainers noted in response to the question 'Why women-only?': "The theory is, basically, that we [...] build up the skills and the confidence of women who've had bad learning experiences or no learning experiences in a non-competitive environment". Brew and Garavan (1995:17) find that 'women feel more able to admit learning blocks and to tackle issues which might be especially pertinent to them (viz work-home conflicts)' than when men are present. And certainly, the VVS coordinator reported that the kind of women they target are often failed by equivalent mixed-sex courses, because their practical and emotional needs are not supported by these courses.

None of the graduates we spoke to had chosen the course because it was women-only, yet when asked what the women-only training meant in practice, they highlighted many of the positive dynamics noted here. Caroline spoke for many;

> At first I didn't think that was important. But I do know that when you get women and men together, men tend to have more dominant personalities, and they do tend to sort of take over. And I think for a lot of people that were going on the course, there was a sort of lack of confidence, so the last thing you really need is a man in there! [laughs]. [...] I think the women's personalities came out more, and it was easier for everybody to speak.

Trainees most commonly identified the encouragement women provide each other as the advantage of women-only training. This appeared to occur readily once the women got to know one another and realised they are 'all in the same boat'. There is a strong sense of solidarity, and strong veins of humour link the diverse trainees, based in part on difficult life experiences. Many told us they took strength from the other women; Lola, for instance:

> I appreciated the stories of the other women. They all are women that have certain bad experiences, they have divorced, or ... And I find it so impressive that still all of them did it, followed that education and everything.

There does seem to be a prima facia case for a positive role model effect in women-only courses. Many also felt that the course being women-only made it easier to socialise than on a mixed-sex course. In addition, feeling safe in a setting without men may be especially important to specific groups: Muslim women, for instance, and women who have been abused by men.

We found some suggestive evidence that courses like EWTC and VVS challenge a masculine coding of ICT. The women we spoke to from the EWTC course all embraced the instrumental and non-gendered image of ICT conveyed in its marketing – as a set of useful skills that will get them a job – and (when asked) did not see traditional associations between masculinity and technology as relevant in relation to computers (although reportedly some older women do see computers as 'a man thing'). It seems likely that such issues might surface if men were present, however: when the previously women-only Irish CIT course became mixed-sex, women trainees began to 'band together' and position themselves as 'not mathematical' or 'not technical' [MacKeogh 2003b] – in line with the evidence presented earlier about girls 'practicing computer inexpertise'.

There is a logical point here signalled in our introductory discussion – namely, in a single-sex setting there is less scope for gender-differentiated behaviour around technology to emerge. Thus, on a women-only ICT course, with women trainers, it is difficult to sustain the notion that only men can 'do' computers, since trainees are surrounded by women who provide positive role models for becoming computer competent and confident. Thus, in the context of the course itself at least, a masculine coding of ICT slips away. At the same time, it must be acknowledged that there is another factor operating here, quite apart from the positive impact of the women-only course. In the case of the EWTC especially, many of the jobs the women are being trained for and go on to, are strongly gendered as 'women's jobs'. This highlights the irony that masculine codings of ICTs have occurred even though the earliest 'mass' use of computers was in routine office work by women.

In sum, the testimonies of former trainees illustrate just how powerfully women-only vocational ICT training can raise the horizons of women who have been out of the labour market for some time. A clear implication from our research and others (Brew and Garavan 1995:17, Hammond 1986) is that the role of such women-only vocational training is *transitional*; it is a stepping-stone to women then re-entering the mixed-sex world, but on a more equal footing. This finding resonates strongly with the argument from early second wave feminism, that women-focused practices and women-only 'safe spaces' are a necessary prerequisite for gender equality (cf. Bondi and Davidson 2002). The continued reality that huge numbers of women are trapped in a downward spiral of low self-esteem, confidence and goals, reinforcing limited employment opportunities, means that women-only courses on the model presented here offer a tried and tested strategy for improving women's prospects on the labour market (see also Leathwood 2004, Skelton and Becky 2002).

In these types of strategies, the category 'women' is clearly intended as a very specific sub-category of women; the whole package of measures is tailored to meet their particular needs. And as we have seen, this women-centred space is especially effective in building up the confidence and self-esteem of those who lack this. Our analysis indicates three ways in which being 'women-only' contributes significantly to the success of such courses: (i) creating a non-competitive environment which encourages mutually supportive dynamics; (ii) generating positive role model effects within the 'community'; and (iii) nurturing a safe space in which women feel supported in their lives outside and can speak more easily about their circumstances than they would with men around. Significantly, all three elements are present in the other women-centred spaces we studied.

Networks for women in ICT: A safe haven from men?

In Chapter 8, we learnt about networks set up by and for women working in ICT occupations. Like the women-only ICT training courses described early, such networks emerged during the 1980s from an explicit feminist politics. They grew out of the experiences of women who had experienced the 'chilly climate for women' in ICT and other fields of science and engineering. Accordingly, the networks for women in ICT typically had a twofold aim: they sought to fight gender discrimination in ICT occupations and organisations, and to empower women working in the sector to further their careers. Women in Technology International (WITI), one of the oldest and largest of these networks, is typical in this regard. It now has branches across North-America, Australia, Ireland and the UK, operating as a professional self-help group, to support women to fight gender discrimination in the ICT sector. In Chapter 8, we analysed UK and Irish branches of WITI [Pitt 2004b, MacKeogh 2003d], plus a more recently formed UK network called Untold [Pitt 2003b, 2004c] and the virtual WOW network in the Netherlands [van Slooten and Rommes 2003].

Clearly, just like the women-only courses, these networks were targeted on – in this case, created and run by – a very specific sub-category of women: women who work in ICT roles. And here again, the motivation for these women to join is very strong. As we showed in Chapter 8, proportionately fewer women than men in the science and engineering professions stay and get on in these occupations; many find they are overtaken by their men colleagues in terms of promotion and/or end up in career dead ends. Key factors here are long working hours and consequent poor work-life balance; the numerical dominance of men resulting in 'boys networks' and workplace cultures which marginalise women and inhibit their career progression; and the invisibility and stereotyping of women's technical ability, which serves to undermine women's confidence.

The activities of these networks are principally geared to the latter two of these problem areas: career development and visibility. Members engage through both web-based resources and various off-line activities: discussions, training sessions, contact-building opportunities, speaker events and public debates. The training offered aims to help members feel confident and inspired, believe in themselves and see professional goals as attainable – a surprising parallel with the women-only ICT training courses. And here too, the emphasis is on 'soft skills', provided in a safe and supportive environment. Many of these skills – such as goal setting, career strategies or body language – are taken for granted in the workplace, typically learnt informally from peers, senior colleagues or mentors. Yet, because of their relative isolation and invisibility, women in ICT workplaces often have fewer opportunities and less support to acquire these skills, and to develop the confidence to promote themselves effectively. Accordingly, the importance of role models and mentoring – to inspire and to guide – is widely recognised and encouraged in all three networks we studied.

Chapter 8 argues that a significant 'strength of numbers' effect contributes to the attraction and dynamism of these networks to their members, many of whom are employed in workplaces where women are found in only ones and twos. By sharing their experiences and discovering they have much in common, the women come to understand that their problems at work are not the result of individual failings and so can identify ways to improve their situations. Our studies of how users relate to the WITI, WOW and Untold networks all confirm the huge personal benefits of these women-centred spaces in empowering members to further their careers [MacKeogh 2003b, Pitt 2004a and 2004b, Rommes and van Sloten 2004]. Those interviewed all highlighted the mutually supportive, trusting and safe environment the networks provided; the inspiration and guidance of role models encountered; and the very real professional benefits of contacts made by networking in this way – again echoing the key elements we identified in the women-only training courses.

The 'strength of numbers' effect is also relevant for the impact of these networks on the gender coding of ICT. By their very nature, and more directly through the public events they stage, these networks serve to make more visible the sheer numbers of women with specialist ICT skills, so contributing to shifting the presumption of ICT as a men's domain and helping to 'normalise' (cf. Faulkner 2009b) the woman ICT specialist. Both the Untold and the Dutch WOW networks clearly gave visibility to the presence and ability of women with specialist ICT skills, so contributed to challenging a masculine image of ICT and computing, if not 'feminising' it.

In part, these networks for women in ICT were empowering because they addressed mechanisms of exclusion in ICT workplaces – especially women's invisibility, the paucity of women role models, women's exclusion from informal networks, the undermining of their confidence and ensuing lowered professional ambitions. But they are also effective because they enable some very positive dynamics, providing positive circles of inclusion as we observed in Chapter 7. Not only do the women who participate draw strength from one another in these networks in ways which empower them individually; they themselves also actively shape these networks since their organisational ethic is very bottom-up, participatory and egalitarian – what in Chapter 8 is called 'networks as empowerment'.

There are striking resonances between participants' experiences of these networks for women ICT workers and the trainees' experiences of the women-only ICT courses. Surprisingly, given that the former are not economically excluded or marginal women, we still find issues of confidence and limited horizons – apparently more readily addressed through women-centred spaces than in the mixed-sex world outside. And in both cases, the positive dynamics that appear to be critical in empowering women through these spaces concern not only the skills and insights learnt, but also the safety to talk, and the huge value of role models and of mutual support.

There is a sense of a safe haven from men in many of the women's testimonies. WOW still endeavours to maintain a protected space by excluding men, as did most if not all women in technology networks early on. But times appear to be changing in this respect. Participants in both Untold and the UK branch of WITI very deliberately invite men to their meetings – on the grounds that excluding men counters the egalitarian ethic and may exacerbate resentment and resistance to gender change. On the other hand, many women who participate believe that much of what they value in these networks occurs because they are women's spaces, and do not want to lose this.

To understand this ambivalence, one must remember that these networks are players in a gender power struggle, in so much as they all strive to counter the 'old boys' networks in ICT occupations and workplaces. Significantly, however, this is a power struggle in which, ultimately, the women need to win over 'hearts and minds' of the men in their field; they need them as audience, as advocates and as co-activists, if they are to overcome the invisibility of women ICT specialists and the gender imbalances in ICT workplaces (Lee et al. 2010). Here again we see women-centred spaces operating as a necessary stepping stone to greater inclusion and success in the mixed-sex world of ICT work. So whilst these networks are to varying extents a safe haven from men – in line with the 1970s version of category politics – they also have a clear transitional agenda towards gender mainstreaming. The current tentative steps to open some activities to men – and to open a dialogue with men in ICT – may require the creation of more 'gender neutral' spaces.

Women's spaces on the Internet: A virtual room of one's own?

In the women-only training analysed above, women meet physically. By contrast, the Internet offers the possibility of making virtual women-centred spaces – through web-pages, email-lists, chat-rooms, Facebook, and the like. While it is difficult in practice to guarantee that virtual spaces will be populated solely by women (Bromseth 2006), they may still be predominantly used by women and experienced as 'women's' spaces. For our purposes, the defining feature of women-centred spaces on the Internet is that they are designed with women users in mind. They invite women to interact with other women (or men), to enhance their computer skills and extend their use of the Internet, and thus help to get more women included into the Information Society. So what strategies are being used to achieve these outcomes?

Chapter 4 identified three types of digital inclusion facilitated by websites designed for women. One is the *self-inclusion* that results when women are motivated to log on to websites because the content of the sites is of interest to them. This was evident in the case of commercial women's magazines which have been made available on the Internet. Women's magazines are archetypical settings for the design and redesign of conventional femininities. They emerge from a culture of entrenched dualist constructions of gender, with the result that editors' choices about content have a strong tendency to reflect and reproduce stereotyped presumptions about 'what women want'. Whilst this can be faulted, for performing quite traditional constructions of femininity and womanhood, we find it significant that these web magazines all constructed their target audience as 'the modern woman', not women in general. All the web magazines we studied made efforts to decouple the assumed link between technology and men, at least with respect to ICT.

One of the ways they sought to achieve this goal was by a range of measures geared explicitly to increasing women readers' computer literacy. We have called this second type of inclusion *assisted self-learning*, referring to the inclusion work done by the web magazine producers to help their readers acquire ICT skills. The web designers emphasised that the web pages should be simple and easy to navigate. They also created opportunities for their readers to become more familiar with the use of the Internet as a source of information and learning.

What increasingly replaced the efforts of magazine producers to assist readers' ICT learning were various forms of interactive on-line forums. This *socially embedded learning* is the third kind of inclusion strategy we found practiced in all of the women-centred websites studied. From the outset, the four commercial women's web magazines offered options for interactivity through various types of forums, where readers had the opportunity to create their own content and, in the process, to create some kind of on-line community. This effort was clearly effective as a digital inclusion strategy, in the sense that it was appreciated by readers and meant that some of them spent a lot more time on the Internet than they would otherwise have done. It also served to generate content which (some) readers found interesting. In this manner, the web magazines shifted from being mainly suppliers of information to becoming vehicles of web-based communication. From the cyberfeminist perspective, it is to be expected that this shift to interaction and communication would help make the Internet and the Information Society more interesting to and inclusive of women.

Indeed, this reading is at least partly supported by our studies of two informational websites for women emerging from voluntary initiatives. One of these, the *Lupus* website [Fortunati and de Luca 2003, Fortunati 2004b] was established to provide information about a rare skin disease, which mainly affects women and young girls. Because Lupus is so uncommon, the site also became a space in which a virtual community readily developed – a community prevalently of women (sufferers and their mothers), who gave each other support and solidarity. In this way, the *Lupus* website became a place in which to find fellowship as well as information. This created a kind of virtuous circle between the growing body of information and the growing community of users – a dynamic which served to increase the attraction of going on-line to use the *Lupus* website, as well as raising the levels of ICT competence among *Lupus* users considerably [Fortunati 2004].

We propose the term *doubly articulated inclusion* to capture the combined interactivity of content (net-based information) and community (mediated forms of interaction), which is so dramatically illustrated in the case of the *Lupus* website and which is increasingly evident in the social networking technologies like Facebook. This form of inclusion creates a positive circle in the sense that increases in the volume and quality of the website content lead to increases in the volume of participating women interacting with each other through the on-line community; and *vice versa*. We believe this is a powerful mechanism of inclusion precisely because it is dynamic and self-sustaining. And we believe that this doubly articulated inclusion was operating to some degree in all of the websites for women we studied.

The Dutch version of *Women on the Web* or WOW [van Slooten and Rommes 2003, Rommes and van Slooten 2004] was set up to provide and exchange knowledge about the Internet among women, drawing on the expertise of women ICT experts who participated on a voluntary basis. A survey showed that users appreciated the website and the activities offered not simply because the information was useful, but even more because it was given in a respectful way and because it was embedded in supportive community of women who also acted as role models for one another. As one interviewee reported:

> WOW was most relevant for me because of the confidence and moral support that they have given me, and that I needed in order to work independently as an entrepreneur. Sometimes women on the list will write about 'their own little enterprise', and then some of the women will reply 'come on, women, we do not have a small enterprise, we have our own company!'

As we see here, the community interactions in these forums provided very similar benefits, and had very similar dynamics, to the women-only vocational training and to the other women's networks described earlier. All of these women-centred spaces empowered women and built their confidence by enabling mutual support and role modelling, and providing safe spaces in which to talk about one's personal circumstances. And in all of these women-centred spaces, the stronger the community grew, the greater the mutual support and the more effective the digital inclusion. Communities and community building were far more prominent and enduring in these networks than in the women-only training, where the participants only constituted a community for the duration of the course.

As we demonstrated in Chapter 4, the women-centred spaces on the Internet represent a quite successful inclusion strategy. They have played an important role in motivating and supporting self-inclusion and self-learning among users of the websites, as well as engaging women in shaping the technology through their participation in on-line forums. Significant numbers of women have been recruited and socialised into the Information Society by this route. In this way, the strategy has also been successful in transforming the symbolic image of the Internet as 'men's space'. But to what extent and in what ways are these inclusion outcomes dependent on the websites being women-centred spaces? Insomuch as these are (with the exception of WOW) not women-only spaces, this question could be answered in the negative. 20 per cent of the readers of *Donna Moderna*, for example, were men, some of whom participated in the forums [Fortunati 2004a]. What matters here, however, is that all the websites were targeting women, not men. Our analysis indicates that being women-centred has had a direct bearing on their success as inclusion strategies in three ways: first, by providing attractive content to motivate users to go on-line; second, through the effectiveness of the designers' efforts to support their users' journey onto the Internet, by assisting their acquisition of ICT skills; and third, through the creation of spaces which enabled the women users to become active participants, thus shaping the form and content of the websites so facilitating their further inclusion.

Designing for girls or women

Our evidence on the web-based women's magazines indicates that it is possible for inclusion strategies based on category politics to come from commercial actors. In this case, the strategies involved a 'feminisation' of the product by providing content assumed to be attractive to 'modern women'. To what extent is such inclusion by feminisation also feasible through the design of other kinds of ICT products?

We approached this issue in Chapter 6. Two of the companies we studied, drew explicitly upon gender stereotypes in their design efforts. One of them was a UK business that had produced a series of 'girls' games', entertainment CD ROMs for early teenage girls [Stewart 2003]. Their design strategy seemed to follow a well-established tradition within the toys industry, namely to work from a dualist set of gender stereotypes. The company's designs worked from the idea that boys and girls have different preferences in terms of themes and content. In making games exclusively for girls, the company drew upon themes such as the secret diary, the pop-star emulation game and the game of being in love. They also chose a particular aesthetic with very pink, cartoon-like and irregular shapes, taken – in a clear parallel with the commercial web magazines – from widely read girls' magazines:

> The main difference comes from the content, when we make girl games we choose a different content. We build a diary, how to become a star, my Idol, which you collect pictures from the Internet, your favourite movie star, for example.

The company considered this strategy reasonably successful in economic terms; they have found a profitable niche market in an industry that is mainly focused on boys.

The other company which explicitly drew upon stereotypes was Philips and its attempt to design an electronic product for girls aged 7 to 12, called KidCom. They pursued a somewhat different, two-prong strategy to their design endeavour [Rommes et al. 2003]. First, they put a lot of effort into developing a methodology that could provide reliable information about what girls in the chosen age group might be interested in. Second, they tried to reach girls by aiming at a common denominator of 'girlishness'. An important goal of the design process was to define in what ways girls actually 'are' different from boys. In terms of the design strategies described in Chapter 6, Philips combined reliance on stereotypes with an element of user inputs. Thus, gender differences between boys and girls were constructed from a number of sources: designers' own views and experiences, literature, expert advice and feedback designers got from children with whom they tested the concepts. As a result of these inputs, KidCom was designed as a communication device, rather than a game.

A common aspect of these two cases is the choice to base the design strategies on presumed differences between girls and boys. An alternative approach would be to start from a presumption that there is considerable overlap between the tastes and interests of girls and boys, and to provide designs that could be considered gender indistinct. Aphra Kerr's [2004] analysis of game-playing among young Irish women found that when young women are introduced to computer games they find them quite entertaining, even if they are critical of some 'boyish' aspects. Often, the women were introduced to the games in the context of relationships with young men, but the women continued to play such games even if the relationship ended. In the language of this book, we see here that inclusion work was needed in order to recruit and socialise these women to the playing of these computer games. Such inclusion work may be related to the context of play as well as the content of the games (Carr 2005, Hartmann and Klimmt 2006, Jansz et al. 2010).

In line with Kerr's findings, designers in four Norwegian computer games companies claimed that more women are into the playing of computer games than is usually assumed [Gansmo et al. 2003]. Like other companies, they saw girls and women as a potential market for computer games, but unlike others they were critical about making games specifically for women. According to them, those men who were very interested in computer games represented a minority and did not constitute a profitable market. Of course, neither did those women who were definitely uninterested in games. This left a group consisting of the majority of men and a considerable proportion of women whose interests were not that different. In the view of these Norwegian designers, these two groups could be catered for by the same game, if the quality of the game was good.

Thus, as we saw in Chapter 6, the really important challenge for these games designers was to 'find a good story' from which to make a high quality computer game. Their main ambition was to make 'better' games, and they claimed that higher quality was the key to enrolling more women game players. Their approach was based on the idea that a variety of tastes and interests could be catered for within the same game. For example, one could add more women role characters in the game, or offer a greater variety of activities that included 'feminine' as well as 'masculine' connoted options. In this way, the Norwegian designers preferred to design gender inclusive or gender indistinct games – a strategy we would characterise as gender mainstreaming, in contrast to the category politics strategy of designing games specifically for girls or women. Where the UK games company and the KidCom design team at Phillips largely developed their design choices from a stereotypically informed, binary model of gender, the Norwegian companies tended to develop their designs from a more diversity aware model of gender.

Of course, our research demonstrates that both approaches can work. The studies of web magazines for women show that it is possible to tailor a product to an audience of women by thinking from presumed gender differences. Nevertheless, attempting to tailor a product for girls/women can be quite difficult in practice. As the KidCom team's handling of the 'pink look' demonstrates, there is no straightforward, clear-cut way to translate 'femininity' into design criteria, because in many contexts 'femininity' has very complex, ambiguous meanings.

Designing for women only, the category politics approach, is clearly possible; but it is also problematic. Quite apart from the practical difficulties noted above, one runs the risk of reproducing traditional notions of what women or girls prefer and need. Where this happens, ICT is symbolically 'feminised' but in a politically retrograde way. The alternative was signalled in the Norwegian game designers' version of gender mainstreaming, through improved quality and increased sensitivity to gender diversity of preferences. Here again, we therefore have suggestive evidence of a transition from women-centres spaces to spaces which are gender-mixed or even gender-neutral.

Does 'feminisation' work? Critical reflections on women-centred spaces

This chapter has explored features of inclusion strategies explicitly targeted on women. We have looked at quite different efforts to achieve inclusion in this way, all of which we have characterised as involving the making of women-centred spaces. We encountered marginalised women who clearly benefitted from participating in women-only vocational training, tailored to their life situation and their particular inclusion needs and to building their confidence and self-esteem. We met women who are already ICT specialists but who nevertheless were empowered to gain resources to further their careers in ICT through networks run by and for women in ICT, networks featuring very similar dynamics to the women-only training courses even though their members are very different – mutual support and role modelling in a 'safe' environment. We saw how many women have become included into the Internet as a result of various women-centred websites, which attracted them by providing interesting content, facilitated their learning of Internet skills, and provided on-line forums in which they could provide and shape their own content and teach/learn ICT skills. Finally, we looked at efforts to extend the market for ICT products to girls and women through women-centred designs, and contrasted these with 'for everybody including women' designs.

In all cases, digital inclusion did occur to some degree at least – both in terms of increasing the numbers of women using or designing

ICTs, and in terms of shifting the gender coding of the technologies. But does this mean that the cyberfeminist notion that labelling ICT, or particular ICT domains, as 'feminine' is an effective strategy for getting more women included in the Information Society? The main problem with this position lies, as we have repeatedly noted, in using categories such as 'feminine' or 'women' in the singular. We find it significant that none of the women-centred initiatives we analysed had women-in-general as their target group. They all addressed particular sub-categories of 'women' – unemployed and marginalised women, women in ICT occupations, 'modern women', women with a particular illness, young girls, gamers and so on. Thus, the construction of women-centred spaces seems to represent an inclusion strategy that is, at least implicitly, informed by a more diverse conception of gender than the binary standard.

This demonstrates that category politics does not have to be based on all women; indeed, this is unlikely to be a very effective inclusion strategy. Starting from the needs or interests of specific groups of women is likely to be more effective, precisely because it recognises the gender diversity we have stressed repeatedly in this book. Arguably, the making of women-centred spaces is mainly about providing a context in which groups of women find other women like themselves, with whom they may exchange experiences and further shared interests – not because they belong to the large category of 'women', but because they belong to a more specific and identifiable group. This exemplifies how category politics in the area of gender and ICT may be made quite specific, without necessarily mobilising unhelpful gender stereotypes or essentialising 'women'.

Two very important implications flow from this observation. First, the things (content) that motivate particular groups of women to get recruited, need to be quite specific and relevant in order to attract that group – skills to help me get a job, help with my career, information about my illness, lifestyle hints, how to get on at work, what I like to do in my leisure time and so on. Second, the women-centred spaces so created need to involve some building of communities. The interactions within these communities – especially the mutual support, the valuable role models and the safety to talk about oneself – are a large part of what helps to sustain the socialisation of participants into the Information Society. It is precisely this specificity of content, plus the important role of communities built around the content, which form the basis for the mutually reinforcing relationship of content and community we have called doubly articulated inclusion. Arguably the specificity of both content and community is why this powerful dynamic is such a common feature of strategies involving the making of women-centred spaces. It is another example of a positive circle of inclusion, even if it operates differently from the one observed in Chapter 7.

At the same time, it is important to observe how often the strategy of providing women-centred spaces is a transitional precursor to a more gender mainstreaming effort, as anticipated by early second wave feminism. This is most apparent in the cases of women-only training and women's network, enabling participants to progress in the mixed-sex labour market. In the cases of networks and ICT products, we saw hints that gender mainstreaming may be becoming a preferred strategy – or at least that mixed-sex spaces might also be helpful. Clearly, a real risk of category politics is that they mobilise and so reproduce gender stereotypes and essentialisms. Foregrounding ideas about how women 'are' can easily produce a standardised notion of 'woman', which in turn may work to curb diversity by reproducing normative pressures. The risk is reduced when the category is made more specific, like 'modern women' or 'marginalised women', but it does not go away – as we saw in the case of the web magazines appealing to quite stereotyped notions about what interests 'women' and about women's computer literacy.

Thus, the kind of doubly articulated inclusion efforts we have examined in this chapter suggest a paradox. On the one hand, the productive interaction between content and community seems to demand a women-centred substance with respect to relevance and community dynamics, which invites an explicit concern for sub-groups of women as the target audience. On the other hand, the construction of women-centred substance seems too often to be dependent on gender stereotypes, which in turn may mean that the successful interaction of women-centred content and community may conserve gender as an action category in ways that reproduce gender-related prejudices and inequalities. 'Feminisation' of ICT may be more risky than anticipated by cyberfeminists; much of the evidence presented here suggests that a de-gendering of the image of ICT, transgressing gender binaries and stereotypes, is likely to be more effective in the long run.

We are struck by the implication in Daniel Miller (2011) and Sherry Turkle's (2011) recent commentaries on the topic that gender is not an important category of analysis in relation to the use of Facebook and similar media. It may be the case that these social networking technologies have become de-gendered – in the sense of no longer being marked symbolically as either 'feminine' or 'masculine' – in spite of evidence that men and women use Facebook in different ways (Lewis et al. 2008,

Mazman and Usluel 2011). Nevertheless, the gender differences with respect to Internet skills reported in Chapter 1 at least call for some caution. In any case, the popularity of Facebook and the like confirms the importance of doubly articulated inclusion, since the interplay between content and community is an obvious source of the success of these media (cf. Miller 2011). Not only the dynamic of the new social media as an inclusion mechanism, but also the skill set needed to engage with this set of ICTs, are essentially an extension of what we have observed in the chat rooms and other web-based interactive sites described in this book.

Without doubt, however, the gender coding of ICTs remains a pertinent and difficult issue. In the next chapter, we explore this problem, and the paradox of doubly articulated inclusion strategies, further. In particular, we analyse how gender coding is produced, and how this is related to the relative numbers of men and women present in any area of ICT.

10

The anatomy of inclusion

Knut H. Sørensen

In previous chapters, we analysed different types of efforts to include more women into the Information Society, as users and/or designers of ICTs. Here, we want to summarise these findings in order to offer a more systematic approach to the understanding and construction of inclusion initiatives. To do so, we seek to describe what we have chosen to call *the anatomy of inclusion.* The anatomy metaphor is meant to emphasise the need to map and classify the underlying strategic elements of inclusion initiatives, or rather, the underlying *inclusion strategies.* In Chapter 1, we defined an inclusion strategy as an activity aimed to (1) recruit people into and/or (2) retain and socialise them within some system, in our case women into the Information Society. Some of the efforts we studied were not specifically aiming to get more women into ICT, but they did nevertheless succeed in including significant numbers of women. They have, therefore, all been part of our analysis, and all efforts provide lessons that give important insights into the dynamics of inclusion processes and strategies.

So what are digital inclusion strategies made of? The efforts we have analysed display considerable diversity, along many dimensions. This reflects the variety of situations and problems addressed as well as of the resources available to achieve inclusion in a given context. In particular, we have found that the following three elements of our anatomy of inclusion, taken together, provide a useful framework for thinking about the ways different strategies are made and how they work:

1. How and why activities and actions are initiated.
2. The underlying strategy: understanding of the inclusion problem(s) and the possibilities for change.
3. The choice of inclusion instruments and actions.

The chapter is structured around these three main elements of the anatomy of inclusion strategies. We begin with the initiation of inclusion initiatives, addressing the various motives behind the digital inclusion initiatives we studied, and exploring the role of feminist subpolitics alongside more institutional drivers. Turning to strategic thinking about gender inclusion in the Information Society, we pick up a theme which runs through our whole study – understandings of gender and the challenge of gender stereotypes. We also revisit the relationship between removing exclusion and developing inclusion, signalled in Chapter 2; how the problem of including women into the Information Society usually has been understood from an exclusion perspective, meaning that inclusion is thought to be about removing or remedying the excluding mechanisms or practices. Whilst acknowledging that some inclusion efforts have built upon a combination of inclusion and exclusion thinking, we highlight the strategic merits of an inclusion perspective, in particular the merit of classifying types of inclusion instruments accordingly the four inclusion needs identified in Chapter 3. Continuing our reflections on the choice of instruments, we also highlight the merit of combining inclusion instruments in ways that achieve 'positive circles of inclusion'.

The chapter ends by discussing some overarching inclusion principles. We offer some cross-cutting arguments about what we call quantity and quality. And we conclude by outlining some important examples of what we are calling 'technologies of inclusion'. This idea draws on concepts such as political technologies or technologies of democracy (Callon et al. 2009), which in turn are inspired by Foucault's work on discipline and power (Foucault 2007). The main point is to observe how approaches to inclusion challenges may be standardised and thus contribute to more stable ways of performing such activities. The concept also recognises how standardised approaches may be particularly effective because they may draw on discursive power emerging from their accessibility and stability. We think that our studies have identified some technologies of inclusion in this sense, and we want to describe them in order to make them more visible and accessible to others.

Initiating inclusion

What are the motives underlying inclusion initiatives, and who initiates them? We started out by studying three main categories of initiators: public sector, private sector and voluntary sector actors and found interesting examples of initiatives in all of them. Not surprisingly, the examples reflected a variety of motives as well as ways of establishing inclusion initiatives. In some cases, the set up depended on enthusiasts, people with ideas and the resourcefulness needed to put them into action. But we also found that inclusion could be a professional activity, a normal part of someone's work. In addition, there were different sources of motivation. In many cases, it had a political origin – for example, in a feminist standpoint or in welfare state politics – but enthusiasm could also emerge from commercial interests or a wish to combine political motives with the pursuit of business. Was this diversity consequential, and if so, how?

At the outset, we expected that governmental concerns about a gendered digital divide would play a major role as a source of activities aimed at inclusion. Actually, in the cases we studied, it did not. Norway was the only exception, maybe due to its long-standing tradition of governmental efforts to promote women-friendly policies, so-called state feminism (Gansmo et al. 2003b, see also Bacchi 1996). Here, we studied how a state feminist mode of thought inspired an education policy explicitly concerned with the problem of 'girls and computing', albeit with limited success due to local autonomy with respect to curricular and extra-curricular practices [Gansmo 2003a, b]. In other countries, local and regional authorities generally appeared to be rather more important than central government, in particular with regard to initiatives bearing on education and training, although these initiatives were in most cases not targeted specifically on women. Public sector efforts to enhance digital inclusion were often prompted by concerns related to the labour market and social policy. But as we learnt in Chapter 3, some also sought to empower their audiences with respect to everyday life and community activities, such as keeping in touch with family and friends or pursuing one's hobbies.

Inclusion strategies initiated outwith the public sector tended to be motivated by what we could call feminist 'sub-politics' rather than by any kind of state feminism. Ulrich Beck (1997) introduces the concept of sub-politics to describe what he sees as a shift from formal governmental activities of implementation (politics) to initiatives emerging out of everyday life – 'social arrangement from below' (Beck 1997, p. 104). To be engaged in sub-politics means that citizens take their concerns into their own hands. Thus, feminist sub-politics contrasts to state feminism, where government is expected to act and to make other institutions act. The Dutch example of the Women's Square [Rommes 2003b] illustrates the point nicely. Here, it was women's organisations that constructed a presence in the so-called digital city of Amsterdam to achieve inclusion of women, not the formal organisers whose stated mission was inclusion 'for all'. The analysis of women's networks in Chapter 8 showed how such initiatives 'from below' – all made by women – represent feminist sub-politics, albeit with differing kinds of feminist politics. For example, the Untold network [Pitt 2003b, Pitt 2004c] was set up by women digital designers to improve their working conditions and to construct a space for exhibiting their work in order to increase their visibility. Both men and women could be members. By contrast, the Women In Technology International (WITI) networks were set up exclusively for women to promote women's careers in technology [MacKeogh 2003d, van Slooten and Rommes 2003].

The private sector may be influenced by feminist sub-politics, as we saw in the engagement of feminist entrepreneurs (Cassell and Jenkins 1998b, Spilker and Sørensen 2002). Commercial interests may be combined with feminist goals – for example, through efforts to design ICT intended to cater to women's needs and tastes – even though we know from Chapter 6 that this is difficult. To some extent, the web magazines for women we studied sought to pursue feminist goals. For most ICT products, however, the interest in targeting women was driven more by the desire to open up new business opportunities. This is evident from several of our studies. Stewart's [2004] analysis of a company providing games for girls found that this interest had emerged from interactions with the market: it was mainly an issue of new technology paving the way for new business. Similarly, the study of Phillips' development of Kidcom showed how the idea to make an electronic game for girls was driven by market and technological considerations, rather than political ones [Rommes et al. 2003].

Inclusion efforts may also emerge from the professional interests of those people employed to do inclusion projects. Training initiatives and local resource centres exemplify the ways digital inclusion efforts may be institutionalised. Similarly, the editors of the web magazines were professionals, as well as the designers of computer games and other ICT products. In many ways, a professionalisation is beneficial because it makes initiatives more stable and less dependent on voluntary or spontaneous engagement. However, a recurring problem with local public sector projects is that they suffer from 'funding blight'; they are only ever funded on a short term basis and even good projects get abandoned when other initiatives become fashionable. Still, they are carried out by people who may continue to operate as 'professional includers', and who build up capability in this through their careers.

The variety of aims which bring people to do inclusion work means that it is possible to establish initiatives based on a diversity of motives and, thus, to engage with a variety of interests in order to get support. Of course, some instances of inclusion of women have been piggy-backed on wider efforts to increase the general market for ICTs. Perhaps the most striking example of this is the mobile phone, which now has about as many women as men users (see Chapter 1). To begin with, mobile phones were marketed towards professionals – typically men – but this has clearly changed. Arguably, this particular ICT provides evidence of the force of diffusion (Rogers 2003). From this perspective, women may be laggards but eventually they come on board. However, as Nordli and Sørensen [2004] show, men and women tended to have different trajectories of becoming users of mobile phones. For many women, this trajectory involved what we have called assisted self-inclusion (see Chapter 4): they were persuaded to buy or were given phones in order to be available for calls or SMS-messages from family, friends and partners.

Inclusion nearly always involves work – a point that tends to be overlooked in the diffusion literature (Rogers 2003). Inclusion work may be done by a variety of actors, not necessarily those involved in the design and marketing of products. Moreover, inclusion work is not always embedded in feminist politics but may be motivated in quite diverse ways. For example, the Norwegian mobile phone study shows how the support given by family and friends to facilitate assisted self-inclusion often was motivated by their interest in making the person more accessible to them [Nordli and Sørensen 2004]. The cyberfeminist argument (e.g., Plant 1997, 2000) that ICTs are particularly suited for women and consequently will be taken up by them, overlooks the work needed for inclusion to happen as well as the diversity of inclusion efforts in terms actors involved, motives and strategies.

There is no simple formula for understanding how inclusion initiatives come about. However, we have observed repeatedly how important it is that such initiatives are deeply rooted in the organisation or institution that is expected to support them, as with the Women in Computing Initiative (see Chapter 7). Further, it may not be sufficient to embed the idea at the top level only. One of the reasons why the Norwegian government largely failed in its effort to make schools develop concrete inclusion initiatives too get more girls into computing was that individual schools were given too much discretion to shape such actions. Many schools chose not to focus on gendered inclusion or did so half-heartedly [Gansmo 2003b].

A considerable proportion of the initiatives we studied were 'for everyone', without any articulated feminist motivation. As a result, they generally overlooked the possible performance of gender in relation to inclusion. In the cases where state feminism or feminist sub-politics were important in initiating the inclusion efforts, it is interesting to ask what kind of feminism we observed, or rather, how gender was understood in these initiatives. We have mostly used 'feminist' in a broad sense, to designate an interest in improving the situation of women, grounded in a perception of women as needing to become better included in the Information Society. This does not mean that the inclusion initiatives have been clearly shaped by feminist theory – some are, but most are not. To what extent did we find influences of the kind of feminist positions discussed briefly in Chapter 1?

Strategic thinking: The challenge of gender stereotypes

The inclusion initiatives we studied varied considerably in terms of the underlying understanding of gender. Often, this was most clearly articulated through the choice of what groups of people to target. Not all inclusion projects were directed solely at women. Indeed, as noted above, quite a number of actions meant to benefit women sought to do so by including men as well, and many did not find it fruitful or necessary to differ between men and women. Other initiatives aimed at particular subgroups of women or at particular groups of men and women. Thus, we encountered four main types of targeting:

1. All women.
2. Particular groups of women.
3. Particular groups, including women.
4. Everyone, including women.

What understandings of gender were at work in these targeting processes, and how did this shape the inclusion efforts?

We would expect strategies aiming at all women usually to reflect the idea that women as a social category have common features or face similar challenges that must be addressed in order to achieve inclusion. They might be thought to share a lack of information about or insight into the importance and rewarding aspects of using ICT (as in the IT Beat initiative presented in Chapter 5); to be troubled by the frequent presentation of a symbolic unity between ICT and masculinity (as in the Women and Computing Initiative presented in Chapter 7); or to be hampered by the lack of suitable role models (as in the networks for women in ICT presented in Chapter 8). In addition, the choice could be grounded in explicit feminist goals to promote access to women to use a particular ICT, as happened when women's organisations took the initiative to shape the so-called digital city of Amsterdam by establishing a Women's Square [Rommes 2003b].

Actually, it is uncommon for digital inclusion strategies to target 'all women' in a strict sense. In the examples we studied, 'all women' tended to mean 'all young women' or 'all girls'. This did not mean that the targeting was shaped by thorough reflection about what might make young women different from other age groups, but rather that the circumstances addressed by the inclusion strategy were linked to age – such as schools, recruitment of students, etc. When we still think about this as a kind of 'all women' targeting, it is because the strategies deal with a cross-section of young women without articulating concerns which may differentiate that group. This is what we have called inclusion targeted on particular groups of women.

Strategies aimed at particular groups of women were often motivated by social policy or equality concerns, for example, that access to ICTs and the chance to learn to use these technologies would be more difficult for less privileged women. Such initiatives often involved efforts to create spaces for women only, as we observed in Chapter 9, usually based on the idea that women feel more comfortable and gain more confidence in an all-women environment. However, social policy motivated inclusion strategies were more commonly targeted on particular, usually less privileged, groups of both women and men The instruments used tended to differ somewhat when they were not directed solely at women. Nevertheless, efforts to empower participants and to help them find ICT relevant and useful in their lives were present regardless of the sex composition of the target group.

Where inclusion strategies aimed at everybody including women, the fourth category, the thinking of the actors involved often did not address explicitly the inclusion of women. As we have seen, however, 'for everybody' initiatives are important and deserve further scrutiny, for two reasons. First, in the event, some of these strategies recruited mainly women (as in the Ardmore Network initiative). Second, what seems to be an idea of gender neutrality may reflect more complex understandings. This is well illustrated by the marketing of mobile phones, which has been neither gender specific nor gender neutral. Rather, the marketing strategies appear to be hybrids in terms of gendered references used. Some messages play on 'masculine' interests and activities, others on 'feminine' ones, while much advertising cannot be coded meaningfully in terms of gender [Nordli and Sørensen 2003]. A similar phenomenon was observed from investigating computer games [Gansmo et al. 2003, Kerr 2004]. While the inclusion strategies adopted were definitely aimed at everybody including women, the designers we interviewed did consider women's interests and needs when they tried to make new and better games. Also, the Gathering study [Nordli 2003] found that while this large computer party was above all meant as an invitation to young people in general to be enthusiastic about the use of computers, the organisers did have a special concern for young women and tried in a modest way to support them, for example, by providing role models.

So how was gender understood in the inclusion initiatives? In nearly all of our studies, the shaping of digital inclusion strategies was influenced by gender stereotypes, often quite traditional ones. This was due to the phenomenon that we have noted frequently in this book: the widespread tendency to talk about gender in dualised terms – as 'feminine' versus 'masculine' ways of thinking and doing – with all the essentialist and stereotypical connotations this tends to bring. We return to this issue in the next chapter. What we want to note here is that binary thinking about gender affected the choice and construction of inclusion instruments in a more pronounced way than it did the choice of who to target. How did this happen?

Of the cases we studied, the Women and Computing Initiative probably articulated stereotypical thinking in the most outspoken manner, by making gender stereotypes an upfront part of some of its inclusion instruments. Most prominently, the advertising campaigns run as part of the initiative to recruit more women to become students of computer science and engineering very explicitly played on gender stereotypes, drawing on a traditional understanding of femininity to argue that the field was particularly well-suited for women. Central to this argument was a claim that women had better people skills and a more realistic attitude to problem-solving than men.

Stereotypes of women clearly shaped inclusion instruments in many other inclusion initiatives we studied, especially the web magazines for women and the designing of computer games for girls. The problems with invoking dualist understandings of gender become particularly pertinent when such understandings are used as a basis for design. The Kidcom project is a good example [Rommes et al. 2003]. The intention was to develop an electronic game specifically for girls. This raised the question: how are girls different from boys? To address this question, the designers developed a notion of 'girlishness', which they assumed or hoped would be representative of all girls. The final product drew heavily on stereotypes, resulting in a product that was supposed to help children to experiment with behaviours stereotypically associated with women – communication, getting to know oneself, and developing personal relations. As a final stereotypical touch, it was given a rounded shape with a pinkish colour. In this way, the inclusion effort runs the risk of conserving gendered practices rather than going beyond them.

The problem with employing stereotypes may be highlighted by comparing our two studies about computer games, discussed in Chapter 6. The Dutch company analysed by Stewart [2003] designed games explicitly for girls, based on the idea that girls and boys have different interests and play styles, including different patterns of using computers and video game entertainment. The designs drew on a stereotype of girls, partly constructed through one woman designer's reminiscence of her own girlhood. In contrast, the Norwegian computer game designers we studied [Gansmo et al. 2003] were sceptical of the idea of developing games specifically for girls or women. Their thinking tended to transgress any simple dualisms, arguing that by making better games, possibly with more options in terms of activities, they could cater for a greater variety of tastes and interests without labelling them 'feminine' or 'masculine'.

What is at stake here is the inclusion work required in order to address the inclusion need of motivation through the construction of relevance. What kinds of argument are used to make ICT – whether generic or specific – appear interesting, appealing or useful to an audience of (some) women? By invoking stereotypes, audiences are made (as if) standard, often in line with a gender binary. Of course, the use of stereotypes to make a kind of ICT appear relevant and appealing can be effective, as we saw with the girly games made by the Dutch company analysed by Stewart [2003]. However, the strategy also seems to circumscribe users to act according to the stereotype: they should use ICT but otherwise remain within the traditional standards of being girls/women, and the way they use ICT should be consistent with these standards.

We return to this issue in the next chapter, and consider what bearing ICT has on the doing of gender. Our key point here is that the use of stereotypes is a clear but problematic 'temptation' (cf. Haraway 1991) in the construction of digital inclusion strategies. Several of our studies show that invoking stereotypes may be rewarding in so much as it is appealing to, and so does attract, users. Women do read web magazines. Parents of girls do buy girly games. When Bacchi (1996) makes a case for category politics, which often is based on some kind of stereotype, she sees this as a tactical move – in principle, it is wrong; in practice, it works.

We disagree with the latter claim on two counts. First, the use of stereotypes is short-sighted. There may be initial success, but at the cost of reinforcing stereotypes one would otherwise have done away with – precisely because stereotypes tend to underpin gender binaries and thus uphold a system of unwanted gender differentiations and, therefore potentially, inequality. Second and perhaps more surprising, our studies show that in many cases, the success of an inclusion strategy depends on a transcendence of stereotypes. This is related to the diversity of inclusion needs and the importance of recognising the significance of tailoring inclusion strategies to particular groups and circumstances. When the Women in Computing Initiative started, they thought they had young women in general as their target group and emphasised the stereotype of 'woman the communicator'. However, the women who actually chose to study computer science and engineering did so because they were interested in mathematics and science, and attracted by the prospect of a good career. We believe that the marketing of computer science and engineering to young women would have been more effective had it focused on such qualities – in much the same way that the networks for women in ICT professions analysed in Chapter 8 undertook a lot of effort to find out what needs to accommodate. Recent developments with respect to computer games seem to support a similar conclusion, since the trend of making games for girls seems to have run out of steam; games that transcend gender stereotypes appear to be more successful (cf. Kafai et al. 2008). This fits with research showing that gender differences in gaming preferences depend on context and mostly result from differences in gaming experience (Jenson and Castell 2011, Jenson and Castell 2008, Carr 2007, Rommes et al. 2010).

What does this evidence tell us about the implications of thinking from category politics versus thinking from gender mainstreaming? The lesson from the making of women-centred spaces is that a move from targeting women in general to targeting specific sub-groups of women makes category politics less problematic. When targeting specific sub-groups one actually needs more knowledge about how to reach them, thus moving from the use of stereotypes to more fact-like information grounded in knowledge of the real lives of the group targeted. This effect is not so prominent when the sub-group is fairly large, such as 'modern women', but more so in, for example, the running of computer courses for less privileged, unemployed women, as we saw in Chapter 9. Knowledge is an important antidote to stereotyping.

Several of the category politics initiatives we studied included measures to ensure lasting effects, by helping the target groups move into the mainstream, and by ensuring that the mainstream caters for more diverse interests and characteristics. Some of the networks studied in Chapter 8 sought to enhance the visibility of the design work of women in the wider industry. Some of the disadvantaged women training initiatives discussed in Chapter 9 made efforts to ensure their trainees were well received by IT businesses they hoped to work. In such ways, these initiatives combined category and mainstreaming politics, deepening their transformative potential for their target groups.

This combining of category politics with gender mainstreaming could be called a politics of diversity. It avoids the pitfalls of category politics because the mainstreaming efforts involved speak to and cater for specific forms of diversity and inequality without interpreting inequalities in terms of gender dichotomies. Inclusion strategies based on such a politics of diversity approach may be particularly effective in addressing the kind of targeting we are calling 'for everybody including women'.

None of the inclusion initiatives we studied actually referred to 'category politics' or 'gender mainstreaming'. They might have benefitted from doing so because these concepts invite reflections about how one does targeting and, as the analysis in the book demonstrates, help to develop more effective inclusion strategies. With respect to networks and ICT products, we concluded in Chapter 9 that gender mainstreaming might become a preferred strategy in time – or at least that mixed-sex spaces and products can play a useful role in achieving digital inclusion, as has also been observed in the case of new social media (Lewis et al. 2008). Certainly, gender mainstreaming facilitates a view of gender that is more heterogeneous and transgressive with respect to binary perceptions than category politics allows. But it also involves risks.

In addition to category politics and gender mainstreaming, we have observed a third kind of gender politics with regard to inclusion: that of 'gender indifference' or 'gender blindness'. Some inclusion initiatives overlooked the potential importance of gender because of tradition, lack of reflection or neglect. This does not necessarily produce failure, but we observed that gender indifference became a problem when one failed to recognise that a given measure is perceived as either a men's or a women's issue or failed to take account of significant factors affecting how (particular groups of) women or men may or may not engage with ICT. We saw this in Chapter 4, with some computer magazines overlooking that men overwhelmingly dominated the readership, and in Chapter 3, with the community initiative apparently uninfluenced by the fact they did not reach many men.

Inclusion models and inclusion instruments

Moving on, how may we characterise the construction of inclusion models and strategies with respect to the choice of inclusion instruments? How were exclusion and inclusion thinking managed in this regard? We have repeatedly in this book juxtaposed thinking from inclusion and exclusion. While our main argument is that an inclusion perspective produces a different outlook on gender and ICT than a concern for exclusion, we found that many inclusion initiatives were influenced by thinking from exclusion or a combination of both. Indeed, some studies have argued for the need to tackle exclusive practices as well as building more inclusive environments (e.g., Bilimoria et al. 2008, Margolis and Fisher 2002, Teague 2000). We should bear in mind, therefore, that there is not always a clear-cut difference between the two mindsets; they are not mutually exclusive.

In Chapter 3, we proposed the concept of inclusion needs. The idea was to identify the main elements of digital inclusion processes that need to be addressed if inclusion is to be achieved. From domestication theory, which focuses on the strategies people employ to become users of a technology in a way they find productive and meaningful, we derived three inclusion needs: those of (i) access, (ii) motivation, and (iii) capability. From the work of Bakardijeva (2005) and Stewart (2002, 2007) we observed an additional category of inclusion needs: (iv) support.

Digital exclusion may be understood as resulting from a failure to meet these needs. The literature on digital divides has focused heavily on lack of access to ICTs (e.g., Norris 2001). Yet, in the examples we studied digital exclusion was more often conceptualised in a somewhat different fashion, drawing on one or more of the following models of the problem:

- *Deficit models.* This way of thinking was perhaps the most widespread. Women's exclusion from ICT was understood in terms of women lacking something. The main deficits were thought to be lack of information about ICT and the benefits of engaging with such technologies, lack of resources to get access, lack of skills, or lack of networks providing support. From this way of defining the problem, the strategy of inclusion becomes a search of ways to make up the deficits, for example by providing information about the benefits of ICT, supplying access to computers or the Internet, offering training or giving support. The women-centred spaces described in Chapter 9 are good examples.
- *Discrimination models.* From this perspective, exclusion happens because women are discriminated against by men, explicitly or implicitly. Such models produce inclusion initiatives seeking to increase consciousness with regard to discriminatory acts. We saw this, for example, in the support networks and other women-centred spaces.
- *Negative image models.* This perspective argues that exclusion is produced because images of ICTs and of using or working with ICT make women reject the technology. Such images are often produced through symbolic links between masculinity and ICT, making the design and use of ICT feel gender-inauthentic to women (cf Faulkner 2009b). The inclusion strategies from our studies that addressed the image problem either aimed for a symbolic redefinition (as in the Women and Computing Initiative), or sought to design ICTs from 'feminine' interests and tastes (as in KidCom or IT Beat).

In such ways, exclusion models were translated into inclusion strategies. Deficits need to be amended or corrected, discrimination countered and abolished, and symbolic links between masculinity and ICT challenged. It would be a mistake to deny that inclusion may be achieved by countering exclusion. However, our studies suggest that strategies which meet inclusion needs are likely to give better results. The Women and Computing Initiative, analysed in Chapter 7, provides an interesting illustration. While the initiative employed several instruments shaped to counter exclusion, such as the advertising campaign, the accounts of the women computer science students emphasised the importance of feeling included rather than feeling not excluded.

We have analysed different inclusion instruments in terms of how and how well they meet the four inclusion needs mentioned above. Although this is complicated, because some inclusion instruments cater to more than one need, a reasonable if not exhaustive summary is presented in Table 10.1.

Table 10.1. Classification of inclusion instruments according to inclusion needs.

Access	Motivation	Capability	Support
Resource centres Increase number of computers at school Quotas Women-centred or women-only spaces	Advertising to link ICTs to women's needs Role models Attractive/useful content Attractive/useful design Attractive/useful Demonstrations which transgress tool/toy binary outcomes of using, learning about or studying ICT Professional visibility	Training Assisted self-inclusion	Networks for women in ICT Mentors

Concern about lack of access to computers and the Internet used to be the main response to digital divide problems. As we learnt in Chapter 1, in the Western world this issue seems less pressing now, perhaps with the exception of the elderly (and particularly elderly women). Computers and Internet connections have become fairly common, mobile phones also. This means that the gendered digital divide is more a matter of actual use – in addition, of course, to the relative lack of women in the computing profession and ICT industry. Consequently, we believe that it is above all the ability to meet the need for motivation, the construction of relevance, which is the critical issue in digital inclusion now. It is no accident that motivation-related instruments have a dominant position in nearly all of the inclusion initiatives we studied.

Of course, education and training remain important, and access to ICTs is not just an issue of their presence in the household but also of their actual availability to the women there. Nevertheless, providing incentives to engage with ICT, whether as users or as professionals, is a key challenge. It has been a commonplace to believe that the lack of inclusion of women is due to deficits with respect to knowing about the potential benefits of using ICTs. This may not be completely misleading, but such information is often neither engaging nor convincing in itself. Information has to be considered relevant to have the desired effect. This is why the double articulation of content and community, described in Chapter 9, is so effective and important. When appealing content also provides the possibility of access to peers, to others in similar situations, the technology or activity becomes much more attractive. Several of our studies have shown the positive effects when women experience such double gains from learning about and using ICTs – whether it be for fun and entertainment, for instrumental achievements in terms of labour market opportunities, for information or visibility, or for social networking. The attraction of new social media such as Facebook seems exactly to be produced through positive circles of content and community. Content is important in itself, but is shaped by concerns regarding the interaction with friends and acquaintances, while the access to such a community make the use of social network technologies pleasurable (cf. Miller 2011).

When considering our four inclusion needs more broadly, it is important to note that there is no straightforward way to translate inclusion needs into design criteria. It was in the course of such translation processes that we saw most clearly the effect of gender stereotypes: relevance and appropriateness of ICTs were often constructed on the basis of stereotypical thinking about women's needs and interests. With regard to design as well as incentives to use ICTs, it is particularly important to overcome the binary between perceiving women as serious users of ICT tools and as engaging playfully with ICTs, as toys. This dichotomy often works to make women appear dutiful and earnest with respect to technology, leaving to men to be enthusiastic and experimenting. A number of the inclusion initiatives met the inclusion need of motivation by inviting women to have fun with ICT, but with the subtext that ICT could also be useful. We believe these efforts of transgressing the tool/toy dichotomy were both important and fairly successful.

Many other conventionally gendered dichotomies need to be transgressed, not least, the technology/people (Faulkner 2007) and altruistic/career-oriented binaries. In a way, women in the ICT professions have been haunted by a digital 'good girl' figure which, juxtaposed to the stereotype of the a-social man hacker, has symbolised people-oriented altruism. The message in this 'good girl' image is that women working with ICT should prefer people over machines, be mainly concerned with technology's usefulness, and eager to do helpful things. Thus, it seems that we still need to emphasise that women should have fun as well as being ambitious and pursue careers when they want to.

Most inclusion initiatives employ more than one type of instrument. The most comprehensive example we studied was the Women and Computing Initiative, which tried to meet all four needs. A quota, reducing slightly the entrance requirements for women students, was introduced. An unexpected effect of the quota was that prospective women applicants interpreted it as evidence that the university really wanted more women in the computer science and engineering programme. In addition, the women students were given exclusive access to a dedicated computer lab. The initiative advertised in mainstream media, they invited all accepted women students to an information/demonstration day, they tried to make role models available, and they emphasised throughout the attractive outcomes of a computer science and engineering education. Some (though not many) curricular reforms were made, and the women students received extra support to build good relations and networks among themselves. In this manner, the Women and Computing Initiative sought to improve socialisation and retention as well as recruitment.

We concluded that the success of the Women and Computing Initiative depended in large measure on this wide spread of instruments, which helped to create what Lagesen in Chapter 7 calls positive circles of inclusion – a situation where instruments are mutually reinforcing. This is a very important concept, to which we return in the concluding section of this chapter. Of course, comprehensiveness is not a requirement for success, and more modest and less broad initiatives also worked. But it may be significant that such a big range of measure as we saw in the Women and Computing Initiative was necessary in order to get more women into computer science and engineering. This, after all, has proved to be a hard nut to crack, arguably the hardest nut to crack in terms of women's inclusion into the Information Society.

Quantity and quality

Cutting across the digital inclusion strategies we have studied, two further issues warrant particular attention. One is related to the number of women included as a result of the strategy; the other is about the quality of what is offered through the inclusion effort. Let us start with the quantitative dimension. At the heart of every inclusion initiative lies some ambition to increase the numbers of women using or designing ICTs – as measured in how many use a particular ICT, participate in ICT-related activities, acquire particular ICT skills or competences, and so on. These quantitative aspects of an inclusion strategy have a dual property. Most obviously, they are how we measure the outcome of an initiative, its success or lack of success. At the same time, we would argue, numbers are at the very heart of the way inclusion initiatives may work. It is the latter feature that we want to elaborate here.

The number of women participating in a given Information Society practice has two main effects which may be addressed through inclusion efforts. The first is related to the problems of being a minority. Kanter (1977) convincingly identified many negative minority effects. When women are a minority, they tend to feel marginal, like outsiders, not normal. They experience problems related to visibility and access to important networks which, in turn, blocks inclusion and may produce exclusion.

There are ways to tackle minority problems. The analysis of our studies has demonstrated to us that most important amongst these is the creation of sub-spaces with a different gender balance, what we have called women-centred spaces. This process is exemplified by networks for women in ICT such as Untold, WITI and WOW. Building such networks is a way of avoiding, or at least reducing, the feeling of being in a minority position. They do this by providing possibilities for women to exchange experience and express mutual engagement to support and sustain each other's efforts as students or professionals in ICT specialisms. Efforts to make teams gender balanced, whether in education or at work, are also important. The management of ICT companies should heed well these clear lessons concerning minority problems.

An example of a successful managerial intervention took place in a small Italian company, Work Up. It was established and run by a woman electronics engineer [Fortunati 2003d]. In this company, 40% of the employees were women, as a result of a conscious inclusion policy originating from the founding manager. She thought women, especially those with experience of raising children, had very useful skills and even hired such women without previous ICT training. They were then trained in the company, which offered Internet services, consulting as well as web development, promotion services and maintenance. A key lesson from Work Up was that characteristics usually perceived as deficits – low mobility, limited flexibility, and family responsibilities – could be turned around to be seen as resources for a successful company.

The second effect of increasing the numbers of women participating in ICT-related practices through inclusion efforts is related to the symbolic interpretation of those practices. Whether people interpret particular ICTs, or particular ICT uses or competences as 'masculine', feminine' or gender neutral is closely linked to the numbers of women and men present, as users, students, specialists, etc. (Lagesen 2007, Faulkner 2001, Sørensen and Berg 1987). When men dominate numerically, this seems to produce a perception of the activity as 'masculine'. When the presence of men and women is more equal, the activity is less prone to being interpreted as gendered. For example, in Malaysia where there are about the same number of women as men studying computer science and engineering, this education is not thought of as 'masculine' (or 'feminine', for that matter) (Lagesen 2008, Mellström 2009). Since a symbolic link between ICT and 'masculinity' may spur exclusion of women, this symbolic property of quantitative relationships – in our case, relative numbers of women and men – is important to inclusion strategies. What may be done about it?

This may look like a kind of chicken-and-egg problem: which should be tackled first – relative numbers of men and women or masculine images? An influential response to this problem has been the theory of 'critical mass' introduced in Chapter 7 (see, e.g., Etzkowitz et al. 2000, Oliver and Marwell 2001). 'Critical mass' designates the number or percentage one needs to reach in order to make inclusion self-sustaining. In relation to gender and science, the most relevant area for our concerns, 'critical mass' theory holds that a threshold level – in terms of the percentage of women in a discipline, department or research group – is needed for minority problems to recede. Once this level is reached, the argument goes, the culture is perceived and felt to be more favourable to women, which in turn means that more women should find it attractive to pursue a scientific career within the department or discipline. The size of the threshold is under dispute. Etzkowitz et al. (2000) argue that about 15–20 per cent women should be sufficient.

Strategies to make an education or an occupation attractive to women by thinking in terms of critical mass and focusing singularly on increasing the number of women are in themselves controversial (Lagesen 2007, Etzkowitz el al. 2000, Grey 2006, Corneliussen forthcoming). Our own view is that directly increasing the relative numbers of women working in or using ICTs is strategically crucial, in a necessary but not sufficient way. Our research leads us to believe strongly in the inclusive potential of a 'strength of numbers' phenomenon (cf. Lagesen 2007). However, it has also taught us that the required strength of numbers may come in the form of symbolically significant representations as well as bodies. Indeed, it is arguably important to work on both fronts simultaneously. Measures such as the consistent use of pictures of active and eager women in all visual representations of ICT work, and the provision of positive role models, may be interpreted as efforts to simulate a more even gender balance through a kind of 'impression management'. And such efforts do make a difference: the more people see women interacting with ICTs, the more normal this appears and feels.

In short, it is vital that digital inclusion initiatives pay attention to the visibility of women as ICT professionals and users. If there are a relatively large number of, for example, women studying computer science and engineering, or using a particular ICT, one must ensure that this fact is noticed. If numbers are less than wished for, one should look for ways to construct a perception that the number of women is greater than it really is. Of course, following the argument above, the latter effort can only work, if it is part of a set of measures capable of producing 'positive circles of inclusion' – measures which would include more direct efforts to recruit and retain women into the Information Society.

Women's networks illustrate well these arguments about the quantitative aspect of inclusion. In effect, they temporarily amend the marginal position of women in ICT workplace and educational settings, by bringing them into a setting with only women or a (vast) majority of women. While there may be a small number of women in any given company, and a relatively small number of women across the ICT industry, when drawn together these small numbers add up to quite a lot of women. Numbers are both relative and absolute; what counts will vary according to the particular context. The supportive and empowering role of women's networks happens because the women discover, through meeting, that they are not alone and that their problems and concerns are shared – the double articulation of community and content. The impact is sustained, at least in part, because of the tangible evidence so provided that there actually are lots of women in ICT. In principle, all types of women-centred ICT spaces have the potential to help change the symbolic marking of ICTs as 'masculine', especially where the numbers of women involved are made visible to the wider world – as we observed in the IT Beat, the women's web magazines and the Women's Day organised by the Women and Computing Initiative.

What about the issue of quality? Why is quality important to inclusion, and what does it mean to be concerned with quality in relation to inclusion measures? Obviously, 'quality' has the potential to mean many different things and to be used in a variety of contexts. Also, the concept may be given normative as well as descriptive interpretations. Nevertheless, a focus on quality catches some important aspects besides the obvious, that high quality initiatives are somehow better than low quality ones. First and foremost, improving quality – as perceived by the targeted audience – can be understood as an important point of departure for developing inclusion strategies. Certainly, improving quality can be an effective way of meeting the inclusion need of motivation – be it through products designed to provide good value, interesting experiences, high levels of utility; there are many ways to phrase it. Among the many actors we interviewed, the Norwegian computer games designers were the most explicit about the issue of quality. They told us they needed high quality stories to make good games, and a high quality story would make the game more attractive to their customers, in general but above all to women. This, they argued, was due to women gamers' particular interest in role playing, where the story is so important to how the game is perceived [Gansmo et al. 2003a].

In women-only training, quality was not just about didactical eminence. Classroom climate and extracurricular support, for example with respect to childcare, were also vital to providing good quality training. Similar concerns were taken on board in the Women in Computing Initiative, not least because the quality of the experience of being a student of computer science and engineering is critical to retention. With respect to the web magazines for women we studied, quality was a factor with respect to aesthetics but also to the attractiveness of content.

The key point here is that, while improved quality benefits everybody, it is likely to be especially beneficial to those who belong to a minority or are on the margins of the Information Society, and consequently are more prone to be in doubt about whether they belong or are good enough. In these situations, improving user-friendliness, increasing attractiveness and providing more effective teaching are all potentially powerful inclusion instruments. We have shown that they all work to get more women included. In terms of gender politics, this could be a win-win situation since men also benefit. On the other hand, quality should not be considered as a disembedded feature. The argument is not that one size fits all, but rather that, under the umbrella of improving quality one may cater for a diversity of needs. With a few reservations, it is tempting to conclude that 'quantity + quality = effective'.

Technologies of inclusion

We introduced the concept of 'technologies of inclusion' in the beginning of the chapter as an ambition to describe generic, potentially standardisable digital inclusion strategies. This proposal flows from our discovery of some ways of doing inclusion which are repeated across the wide set of inclusion initiatives we studied, and which are recognisable as common approaches to the challenges. Our assumption is that, if such technologies of inclusion can be identified and described, this will help the exchange of experience across inclusion initiatives and provide a point of departure for learning how to do inclusion.

One striking feature to emerge from our studies is the lack of exchange of experience, and consequently the lack of learning, between different inclusion efforts. Some professions do have some kind of discourse on digital inclusion – for instance, with respect to education (e.g., Solomon et al. 2003) or design (Ray 2004) – but the initiatives we studied displayed surprising levels of idiosyncrasy. We propose the idea of technologies of inclusion to make planned inclusion work less idiosyncratic. From a scholarly point of view, an additional gain is that the idea of technologies of inclusion invites reflection, following Foucault (2007), as to their disciplining power. To what extent do they work because they make action that results in inclusion, normal, thereby making inclusion the expected outcome?

So far, we have mainly identified what could be considered ingredients in technologies of inclusion. To go one step further, we need to discuss whether there really are aspects of inclusion strategies which could be standardised. Let us start by reminding readers about the anatomy of inclusion strategies. We proposed this could be understood as consisting of the following elements of our anatomy of inclusion:

1. How and why activities and actions are initiated.
2. The underlying strategy: understanding of the inclusion problem(s) and the possibilities for change.
3. The choice of inclusion instruments and actions.

We have identified a number of ways in which each of these elements may be performed. In principle, our observations could be synthesised mechanistically to generate a large number of (strictly hypothetical) potential technologies of inclusion. In practice, many of these possibilities might not work and/or could lack the disciplining power emerging from the public identification of the main strategic ideas as being about the implementation of 'normal' actions. We do not see technologies of inclusion as identical to strategies of inclusion. Rather, we see technologies of inclusion as mainly concerning the third element in the anatomy of inclusion, namely, generic forms of inclusion instruments plus related understandings of how they work and how they should be implemented.

On this basis, we have identified the following technologies of inclusion:

- *Local experts.* This involves the provision of access to people that may be approached for knowledge and support, face-to-face or via electronic networks, for example, in local resource centres or Internet communities. Local experts do not need to be professional computer experts, but they should be experienced in the use of computers in a similar context as the people to be included and have some wider connection with them. (More details in Chapter 3).
- *Assisted self-inclusion.* This inclusion strategy is based on the assumption that women will acquire basic ICT competence if they are motivated and have access to advice, guidance or support to help in their inclusion work. Local experts may facilitate assisted self-inclusion as may net-based information or communities. (More details in Chapter 4).
- *Shared fun with ICT.* Initiatives such as computer parties and web magazines are a way of organising for enthusiasm with respect to computers and other ICTs among young people. They show that these technologies are fun to use and that the use of, for example, computers can be a communal affair involving a lot of socialising with other people. (More details in Chapter 5).
- *Design methodologies.* Through designing with the reflexive I methodology, and especially by introducing user testing or participatory design methods, designers may obtain a more nuanced perspective on gender relations. In turn, this could mean that the use of stereotypes in design possibly would be countered. (More details in Chapter 6).
- *Positive circles of inclusion.* This technology of inclusion is based on the possibility that inclusion instruments may be set up to work in a reinforcing manner, that sets of instruments are selected because they add force to each other. (More details in Chapter 7).
- *Women-centred spaces.* This is a way of generating inclusion, based on two productive ideas. First, to construct a space that provides security and support through communality. Second, to make use of

the dynamic inherent in the co-production of content and community.

- *The double articulation of content and community.* This technology of inclusion centres on the potential of ICT-based activities to provide attractive content and at the same time possibilities for social networking related to this content. (Also described in greater detail in Chapter 9).

These technologies of inclusion are not unambiguously different, but the main underlying ideas are clearly distinguishable. In our opinion, future inclusion work would be well served by developing and using these technologies in a reflexive manner, meaning that ideas and experiences are shared so that learning can happen and practice can improve.

We have not included category politics, gender mainstreaming, quantity or quality in the above list. This is because they represent more overarching ways of thinking about how to do inclusion (the second element of our anatomy of inclusion); they may help to improve and clarify the construction of the various technologies of inclusion, but they are not in themselves technologies of inclusion.

Underpinning the concept of technologies of inclusion is the idea – inspired by Foucault – that such technologies are fuelled by the disciplining power which occurs when ideas or actions are considered to be normal or standard. An example would be that it is normal to use computers or mobile phones to communicate or have fun, which provides a pressure to use these technologies. Did the technologies of inclusion described above provide such disciplining power? We think so. All of them work from an idea that it is normal to use ICTs, perhaps also that ICTs are relevant, productive and even fun technologies for most people. The fact that, these days, it is non-use of ICTs that requires an explanation tells us how much these technologies have become normalised in public consciousness. The exception is inclusion efforts to recruit more women to computer science and engineering; to study computing is not rendered as normal in the sense that there is disciplinary power at work. More work is needed to explore such phenomena (see also Faulkner 2009b).

As argued in Chapter 1, we should also allow for critical views with respect to inclusion into the Information Society. Digital inclusion may not be without downsides. Previous research has argued that the concept of digital divides assumes that the use of ICT is beneficial under all circumstances and for all people. This renders non-use always a problem and neglects the possibility that non-use may be a rational response to particular needs and opportunities (see Norris 2001, Light 2001, Thomas and Wyatt 2000). This point of view was not expressed by any of those we interviewed who engaged in inclusion work. Certainly, in the Ardmore Network (Chapter 3) we observed a case where inclusion efforts were resisted, mainly among middle aged men manual workers. Nevertheless, non-use as an outspoken position was rare in our studies. Commercial actors took non-use of their products as something to be expected. There are many information and communication technologies, and nobody uses all of them. Arguably, however, some technologies, such as the Internet and the mobile phone, are increasingly important tools for people to be able to participate fully in present-day society. We believe that most non-users of these technologies are produced through processes of exclusion, not by informed and voluntary choice.

So what may policy-makers do to improve and further efforts to include women into the Information Society? Some suggestions follow from the discussion in this chapter. First, it is important to create a positive climate for inclusion. This implies that policy-makers publicly recognise the need for initiatives to get women (and all excluded groups) better included in the Information Society, and that they give up the belief that self-sustained diffusion of ICT will solve the problem. It seems from our studies that Norway's tradition of so-called state feminism, of using the government to further women's causes, creates a more positive environment for doing inclusion, not just in the public sector but also in the ICT industry. Policy-makers must strive to show that they take the need for digital inclusion and the 'women and ICT' problem seriously by actively engaging in inclusion work.

Second, inclusion efforts usually demand resources. Thus, policy-makers are challenged to provide more resources. This is not just a matter of money, technology and people. In our concept of resources, we also include knowledge about inclusion. It is important to improve the knowledge base for shaping effective inclusion strategies, and to initiate efforts to make such a knowledge base accessible and relevant to all those who may be engaged in inclusion activities.

Third, as part of improving the knowledge base, policy-makers should facilitate exchanges of experiences and learning about inclusion. This means supporting networking among individuals and institutions engaged in inclusion, and bringing a public focus on successful examples. Much may be gained from a more systematic engagement with positive as well as negative experiences, and from sharing the kind of reflection and learning that we have brought in this book.

Fourth, many policy-makers also need to change their own discourse on gender issues. We have seen in several of our studies how tempting it is to employ quite traditional, dualist gender stereotypes in ways which serve to reproduce gender differences, rather than to transcend them. This means that policy-makers need to engage with and critically reflect on their own use of concepts such as 'men', 'women', 'masculine' and 'feminine' as all-embracing categories. They need to talk about these problems, and acknowledge diversity and heterogeneity, in order to avoid traps such as the widespread presumption that 'one size fits all'.

When is an inclusion measure to include women in the Information Society successful? In our opening chapter, we used statistics to come to a conclusion of qualified optimism regarding some groups of women as users of ICTs. Should we count women users or producers of ICTs before and after an inclusion strategy as measure of success? And can we then say that the increase in numbers was indeed the result of the strategy? This is complicated.

What we regard as success depends on why we think it is relevant to include women in the Information Society. One of the main arguments is that of equal opportunities: for reasons of fairness and democratic rights, men and women should be given the same chances of earning money (wages in the ICT profession are generally higher than in typical 'women's professions') and of influencing ICT applications; and women should have equal opportunities to reap the advantages of being included in the Information Society as users of ICTs. A final argument for including women in the Information Society is for reasons of diversity.

Briefly, the measure of success we propose when studying gender inclusion in the Information Society concern what we might call performances of the acts of inclusion. This would be measured not only by the number of women that got 'included', but also in terms of any changes with respect to gender-technology relations. We have observed a wide range of different outcomes of the inclusion initiatives we studied, which could all constitute measures of success – increased self esteem, changes in the culture of computer science, enhanced possibility of learning, more diverse images of ICT use, to name just a few. What we propose, then, is that transformations in the (ongoing) work in progress of the Information Society might be looked upon as signs of success.

We want to emphasise that the concept of digital inclusion is not just about recruitment, socialisation and retention narrowly conceived. In addition, there is a need to develop inclusion measures that contribute to transformations of both gender and ICT. This is implicit in an understanding of the Information Society as profoundly and simultaneously 'social' and 'technical'. It is also a very clear conclusion of much early analysis of exclusion we have commented upon throughout the book. Whilst some characteristics of the current ICT culture make inclusion difficult, there are also facets of gender – gender norms, gender identities and gendered practices – which arguably need to change if we are to complete the project of reducing the gender digital divide. What have we learnt about the possibilities to achieve such changes? This is the topic of our next chapter.

11

Changing Gender?

Els Rommes, Knut H. Sørensen and Wendy Faulkner

In the previous chapter, we observed how conceptions of gender influenced the shaping of inclusion initiatives and measures. We observed in particular how binary thinking about gender, which often is essentialist and stereotypical, affected the choice and construction of inclusion instruments in a pronounced way. Here, we analyse the making and remaking of gender in greater detail, with a particular focus on the role of ICT.

Sally Wyatt (2008) reminds us about the shifting emphasis on optimism and pessimism with respect to women and technology in the feminist literature. This was reflected in the overview in Chapter 2, where we identified two pessimist types of narratives that we respectively called 'a world without women' and 'a chilly culture for women'. In many ways, such narratives have been predominant in the field, compared to the more optimistic set of approaches, which we labeled 'the woman communicator', including so-called cyberfeminist theories. In this book, we have chosen to focus on inclusion into rather than exclusion from the Information Society. Since inclusion represents a more hopeful take, this means we join in with the optimism rather than the pessimism. However, our optimism is tempered, not unconditional, as indicated already in Chapter 1. We have observed throughout the book that developments with respect to women and ICT are ambiguous. Clearly, more women do become integrated into the Information Society and to a substantial extent in an empowered way. Still, gender differences are being reproduced in many areas, not the least with respect to getting more women into ICT as professionals and designers.

Put in another way, the picture we want to paint on the basis of our studies is not an epic image of progress but rather a struggle – a struggle with the meaning of gender, of ICT and of the resulting socio-technical assemblages. In Chapter 1, we briefly introduced what is currently a common feminist understanding of gender as diverse and relational. Underlying this approach is a critique of the perception of gender as a binary – men and women as opposites – a critique voiced many times throughout the book. On this basis, this chapter discusses two issues which emerge from our research and which have surfaced in previous chapters: the making of gender(s) in inclusion initiatives and meanings of gender in relation to ICT.

The dominant approach to researching gender and technology has been through the idiom of co-construction, the idea that gender and technology are made in relation to each other. In Chapter 2, we criticised the actual studies emanating from this approach, leaning on Lagesen (2005) and Landström (2007), for their asymmetric treatment of gender and technology. Too often, in the analysis, technology is seen as dynamic and malleable, while gender is treated as immutable, stable and given. As a consequence, possible impacts of technological change on gender are hardly investigated. The hesitation to address such questions may be due to a fear of falling back into the trap of the kind of research where technology is treated as stable and women are seen as needing to change, due to (presumed) deficits in skills, competences and interest regarding technology. We do not think inclusion of women into the Information Society is about adapting women to existing practices. So how may we understand gender as dynamic, meaning that we need to study the making of stability as well as change; what aspects should be destabilised and opened for re/de/construction?

Lagesen (2005) proposes two strategies to get around the analytical challenges involved in a symmetrical and dynamic analysis of gender and ICT. One is to study gender and ICT as actively co-produced. Here, she suggests analysing co-production in the way outlined by Jasanoff (2004) – as efforts to make hybrid assemblages stable through the making of identities, institutions, discourses and representations. This idea is useful in our effort to study how gender and ICT were made in the inclusion initiatives we have analysed.

Lagesen's second strategy proposes to investigate more explicitly how gender is affected by ICT, with a focus on scrutinising particular gender-ICT assemblages, or rather, what happens when new ICTs are made part of the hybrid assemblage of gender. The underlying idea is to use actor-network theory (Latour 2005), which focuses on the making of socio-technical hybrids as an approach to analysing various aspects of society. From this perspective, gender would be an achievement, an effect of particular efforts to assemble the diverse human and non-human objects that constitute everyday life. This approach also represents one of several possibilities to rectify the lack of concern for materiality in feminist theory (see Alaimo and Hekman 2008), to pursue Donna Haraway's still relevant agenda outlined in 'A cyborg manifesto' (reprinted in Haraway 1991).

We begin by discussing some features of the co-production of gender and ICT as observed from the various inclusion initiatives, also to assess some effects of the most common ways of doing such co-productions. Then we move to an analysis of how gender could be changed or transformed through encounters with ICT. Finally, we summarise findings with respect to the gender politics of doing inclusion.

Co-productions of gender and ICT: The role of stereotypes

Throughout this book we have looked at various dimensions of gender. We started with what are usually called structural features – meaning the embedded practice of counting and comparing men and women, of studying where and in what kind of communities that men and women are located, what kind of resources they have access to, what they learn and do, and their lived experiences. The relevance of most gender research, including the SIGIS effort, is grounded in the practice of comparing men and women in terms of what resources they have access to, especially observing (statistical) inequalities. Thus, the policy point of departure for this book was observations of numerical predominance of men with respect to the doing of ICTs: in their access to artefacts and practices of use, to training and to ICT related communities, as well as women's underrepresentation in the computing professions.

At the same time, there are problems with such a statistical understanding of gender, above all because these numerical representations tend to make many important aspects invisible, not least, differences amongst women and amongst men. We have further argued that identifying differences between women and men is insufficient as the foundation of policy, since effective reforms also demand insight into the underlying processes which serve to make inequalities. This has been overlooked in some exclusion-oriented projects of the past, which was a main reason why we chose to focus on inclusion: to show what works in practice to challenge inequalities between men and women. Accordingly, we have shown the importance of addressing inclusion needs, such as the need for access, for support and incorporation into communities, or for content which is experienced as relevant in a particular situation. Relevance emerges, for example, from being a young woman, an ICT professional or a person suffering from an illness, needing information about this.

While statistical discourses on gender and technology inequalities permeate public debates, they appeared mainly as a backdrop for the inclusion initiatives we studied, if at all. However, they serve to reinforce the use of gender binaries, which is very widespread in the main public discourses about gender and technology. People tend to think in binaries, such as self as opposed to other, nature versus culture, men versus women, and to make hierarchies out of the binaries, where one part is valued above of the other (Faulkner 2000a). Moreover, it is a widespread phenomenon to gender such dichotomies, seeing one side as 'masculine' and the other as 'feminine'. Clearly, thinking in terms of binaries and simple hierarchies, often resulting in stereotypes, makes life simpler since it becomes easier to classify and label practices. Moreover, it enables one to make assumptions about people without knowing too much about them, which seems to speed up interaction by providing directions about acceptable and proper behaviour (for more examples of what is achieved through the use of stereotypes, see Rommes 2010). Stereotypes and dichotomies even work when we know they are not true: their consequences are still real. They often act as self-fulfilling prophecies as we also noted in the preceding chapter.

In Jasanoff's (2004) terms, stereotypes can be understood as representations as well as discourses, they may provide identity and be the basis of institutions, perhaps be institutions in their own right. The use of stereotypes was prevalent in the inclusion initiatives we have studied, as we also noted in the previous chapter. However, it is striking that it is mainly gender that was stereotyped, not ICT. This seems to mirror the asymmetry of co-construction research, commented upon earlier: gender is rendered stable through stereotyping; ICT is allowed a more dynamic role. This is exactly the downside of deploying gender and other dichotomies and stereotypes: they may become true in their consequences as a self-fulfilling prophecy, and they tend to be reinforced through repetition. The more people believe that women are not competent with respect to ICT, the less confident, and hence less skilled and comfortable, will women be in the relationship to ICT. In this way, the original stereotype is reinforced.

However, when we look more closely at the co-production of gender and technology in the inclusion initiatives, we observe a more nuanced picture. Some of these initiatives mainly reproduced stereotypes, such as the Dutch company that made computer games for girls on the basis of stereotypical assumptions about how girls play. In many cases, however, the effect of working from gender stereotypes was more precarious. For example, with respect to the web magazines for women we analysed in Chapter 4, the content of these magazines, addressing the 'modern woman', was largely shaped by this stereotype. However, some of the magazines added material which did not quite fit the traditional label, such as articles about motoring. Also, the magazines to some extent transformed the image of the 'modern woman' to include computer literacy. Thus, the actual co-production of gender and ICT went beyond the reproduction of the gender stereotype by introducing ICT as a dynamic element in their representation of the women audience.

Similarly, the IT Beat initiative [Pitt 2003a] drew on a stereotype of young women as interested in glam and fashion to provide a link between having ICT skills and being a young woman. Again, the effort to co-produce gender and ICT was an effort to introduce a dynamic element in both: young women belong with ICT; ICT belong with young women. In addition, this co-production provided a demonstration of ICT as an opportunity to be both toy and tool, and of young women as both playful and serious users. In Chapter 5, we saw how effective inclusion efforts could be when they transgressed dichotomies of toy versus tool and play versus instrumental use, dichotomies that sometimes are brought to interact with the gender binary to produce stereotypes of playful men and instrumental women.

An interesting co-production of gender and ICT was observed in the study of Norwegian game designers [Gansmo et al. 2003]. They used degree of interest in computer games as a criterion to open the gender binary, distinguishing between game enthusiastic men and 'normal' men with some interest in game playing, and between women interested in computer games and women not interested. They claimed that in their design, their target group was 'normal' men and women interested in playing computer games, going beyond the idea of computer games as mainly made for enthusiastic men players [Gansmo et al. 2003]. Designing products to cater for stereotypical assumptions of both men's and women's interests, could result in what we in Chapter 6 called 'gender transgressive' products or design strategies. This approach is another way to destabilise gender stereotypes as well as the gender binary.

However, we need to assess outcomes carefully. The widespread use of gender stereotypes in the making of inclusion initiatives is worrisome. Many inclusion actors do not seem to be very conscious or reflexive about their deployment of gender stereotypes. Nor do many see gender stereotypes as inherently problematic, something to be avoided. The fact that the stereotypes in some circumstances may become part of transformative co-productions of gender and ICT should not make us overlook how stereotypes perform to conserve traditional ways of doing gender. Even if the targeted audience of stereotypical messages may claim to be unaffected, even unaware of such meanings (cf. the young women students in Chapter 7), the effects may still be important.

The co-production approach usefully reminds us that discourses and representations with respect to gender and ICT are made; they are not just floating around. Clearly, the inclusion initiatives we have studied did engage in the making of gender, but as we argued above, with somewhat ambiguous effects. On the one hand, fairly traditional stereotypes were mobilised. On the other hand, the involvement of more women with ICT meant some form of transformation, some changes in the making of gender through the inclusion measures. Often, these changes were not radical but they do indicate some level of instability. Of course, the critical next question is to what extent their involvement with ICT also made those women targeted by the initiatives 'do gender' in changed ways?

ICT – a technology of inclusion in itself?

We have in this book not just analysed inclusion initiatives but also ICT-related practices, actions, interactions and related identities. This means that we have studied how people 'do gender' (West and Zimmerman 1987). Throughout the book we have observed the various ways in which gender is performed – for example, by making choices about owning a computer, about how to use it and about studying computer science; by talking about and demonstrating ICT-related skills; by enjoying reading women's magazines on the Internet; or by joining ICT-related communities. What may we learn from analysing such ways of doing gender when we see this as making assemblages of hybrid elements, of human and non-human aspects of everyday life?

Drawing on Latour's (2005) proposal of a theory of action, we could start out by noting that efforts to get women included into the Information Society, to make them designers and users of ICTs, introduce new uncertainties with respect to the making of gender. For example, the identity may shift towards seeing oneself as competent with computers, like the woman who after taking an ICT course concluded that 'I'm not the same person!' (see Chapter 9). When women start to use digital technology, they of course do different things or do things differently than before, like reading a magazine on-line instead of on paper. New objects enter their lives, not just computers, but software, the Internet, mobile phones, wireless connections, etc. When these new objects are used, new relationships are made, to other humans as well as to non-human entities. These uncertainties underpin the capacity for change of the co-productions of gender and ICT in the inclusion initiatives, as we noted in the previous section. They also invite a transgression of traditional ways of doing gender, meaning that traditional ways become strained in some fashion. As we argued in the introduction to this book, a woman with a mobile phone is not the same assemblage as a woman without it. Also, the mobile phone changed when women became users; it was no longer seen as a 'masculine' technology.

Latour argues that the next step in such an analysis is to shift from a concern for uncertainties towards analysing how actors go about stabilising the uncertainties, for example, by establishing new standards. This is a critical point. Uncertainties invite some kind of transformation, but these adjustments or changes need to be made stable in order to achieve lasting effects. What is accomplished by the efforts to include more and more women into the Information Society?

Arguably, doing gender involves engaging with the 'gender cocktail' of various characteristics, choices, actions and expressions that people make around technologies. The inclusion initiatives we have studied provided support for diverse ways of doing gender in relation to technology, such as becoming a woman Internet user, becoming a woman computer professional, becoming a woman mobile phone user, becoming a competent woman participant in new social media, or becoming a woman player of computer games. From one perspective, the inclusion initiatives could be seen to offer new ways of doing gender with ICT, and as we have seen, many initiatives succeeded in making these new ways visible and to some extent stable. The sceptic may ask, so what? How important are the new gender-ICT assemblages to the quality of women's (and men's) lives? Do these achievements represent changes which could be seen as notable progress from a feminist perspective?

There are of course no straightforward answers to these questions, since they depend on the criteria one uses to assess social change as well as one's feminist perspective. However, there is no doubt that there has occurred an increased inclusion of women into the types of activities we designate under the label of the Information Society. In a way, we could say that the number of ICTs present in women's lives in the countries we have studied has increased markedly. To use Haraway's (1991) term, this has made the cyborg quality of the doing of gender more outspoken and visible.

As we have argued throughout the book, inclusion initiatives make more women 'do' ICT. In turn, this of course reduces gender inequalities with respect to ICT, even if some degree of inequality remains (see Chapter 1). In particular, women are still very much underrepresented in the ICT professions. Yet we maintain that there are good reasons for a tempered optimism in terms of the outcomes. For example, we believe that overall, the engagement with ICT that we have studied through the various inclusion initiatives has empowered the women concerned, with respect to this technology and also with respect to wider aspects of women's everyday lives. Still, we have to be concerned with what Haraway (1991) calls the trickster quality of technology. Doing ICT may be beneficial, but it may also be troublesome. The marked contrast between the influx of women as users of ICT and the low, perhaps even sinking, number of women studying computer science and engineering, points to ICT as trickster. ICT is helpful to achieve the wanted inclusion and then it is not, and we cannot be certain when ICT will be the one or the other. The outcome of inclusion actions cannot be predicted with certainty.

It is important to note here what has always been a concern for gender researchers: to what extent do women (or men) have the freedom to 'choose' how to do gender? To what extent are people delimited in their actions by the formatting of their everyday lives by other actors? To what extent are identities imposed by having or not having access to social resources? These are long-standing debates. At the very least, living with ICTs creates some new options for doing gender. Whether these options are made use of is another matter. For example, Clem Herman (2007) and Rommes et al. (2007) show how girls in western countries may repeat dominant discourses about how girls and boys are supposed to be equal partners and have equal opportunities in studying computer science, while at the same time their own choices reflect traditional gender differences.

The complexity of the issue may be illustrated by returning to a point made in the previous chapter: how changing numbers may change the gender symbolism of an institution or activity, for example, by transforming the perception of a profession from being 'masculine' towards a more gender neutral or un-gendered image. As we argued, strong numerical dominance of one gender in a field foregrounds symbolic links between that gender and the artefacts and activities of that field. For example, Oldenziel (1999) observed how engineering became associated with men during the period of early professionalisation in the USA, due to the fact that women were no longer allowed to work in the profession. Conversely, Herman (2007) found how the profession of chemistry scientists in Latvia 'swapped gender' because of a large influx of women.

The research presented in this book has prompted the proposal that there are technologies of inclusion, as presented in the previous chapter, which represent possible building blocks or points of departure for the construction of effective inclusion initiatives. Many readers might have expected also to see an argument that ICTs in themselves may be technologies of inclusion. Actually, the reasoning presented above with respect to the potential of in some way remaking gender – understood as hybrid assemblages – by making one or more types of ICTs part of these assemblages supports such a claim. Most of the inclusion initiatives produced attractive ICT-centred assemblages, such as training, support, useful and/or entertaining content, improved communication possibilities or empowerment.

In this way, ICT worked as a technology of inclusion, but mainly with respect to ICT use. Regarding efforts to get more women into ICT professions, the effects were less clear. The Women and Computing Initiative analysed in Chapter 7 succeeded to some extent in linking computer science skills and attractive career options. But this assemblage depended on a strong labour market for computer science and engineering graduates. Thus, it was vulnerable to labour market fluctuations. In Chapter 8, we showed how professional work with computers was considered in an ambivalent way; such work was not unambiguously attractive. The ICT industry has considerable challenges in making itself more appealing to women.

Can we go beyond gender?

The argument in this chapter so far has been that when ICTs are made increasingly part of women's lives, gender may be transformed. However, such transformation is not unequivocally positive. When we consider transformations of gender in a positive way we are thinking about particular features that should be changed. Obviously, a central gender problem we would like to see disappearing is the production and reproduction of inequalities between men and women. Removing inequality is of course the main rationale of inclusion initiatives, but it is not an easy goal to reach.

Another kind of a positive change would be the dismantling or at least the questioning of binary thinking with respect to gender and the gender-ICT relationship. We believe that the making of increasingly complex gender-ICT assemblages, due to a growing number of ICT elements in the assemblages, will call for greater diversity in gender-ICT representations as well as acknowledgement of this multiplicity. Presumably this will render binary thinking with respect to gender and ICTs in need of (some) reconsideration. In Chapters 5 and 6, we saw how this could happen with respect to the design and use of computer games, when computer games were made part of women's assemblages and women were made part of computer games design assemblages. Another example from Chapter 5 was how young students articulated their navigation between 'cool' use of computers at home and 'boring' use at school by dissolving the toy-tool dichotomy as well as making gender less relevant as an interpretive resource in accounting for computer-related activities. The dismantling of gender binaries, and by implication a deconstruction of gender stereotypes, may in turn lead to a disturbance in or dismantling of some gender hierarchies related to skills and wages – another potential positive change.

An additional gain could be empowerment of women. In several chapters, we have observed how inclusion initiatives increase self-esteem and also may provide a wider set of choices regarding how one wants to live one's life. Again, this constitutes what we perceive as positive effects. Related to this, we would prioritise a doing away with the often unconscious and invisible pressure on people to make them 'do gender'. This pressure acts to limit the freedom to perform identity. The discourses frequently conveyed by many actors about how gender is supposed to be 'done' imply a gender formatting of life that often works to women's disadvantage.

Writing a book like this is, of course, a way of taking part in the making of discourses about gender in general and about gender and ICT more specifically. This has been a reflexive challenge, which means this book is as much an inclusion effort as the inclusion initiatives we have analysed. Accordingly, we should ask similar questions about what this book may contribute to transforming gender as we did with respect to the initiatives, not the least because a book also has performative potential. Actually, this touches upon what could be seen as a major problem facing gender researchers: through studying and writing about gender, criticising inequality, stereotyping and binary representations, we may also call attention to, repeat and reinforce what we are criticising. This is due to the phenomenon Anthony Giddens (1976) called the double hermeneutic circle of social science and the public. While social scientists interpret the public, the public also interprets social scientists. What social scientists report from their studies of the public, performs on the same public. To unveil the harmful features of, for example, gender stereotypes may not be read by the public as exposing the falsity of the stereotypes but rather as a reminder that one may see things the way the particular stereotype invites one to do.

To complicate matters, it is difficult not to lapse into gender binaries when one is writing about gender. This problem is acute in survey-based studies that usually 'measure' gender as a dichotomy: the respondent is either a man or a woman, but it is also prevailing in qualitative analysis. How do we account for gender? Do we want to swap, bend, shift or dislocate gender boundaries? There are numerous examples of how gendered interpretations of professions, artefacts and even colours may be turned up-side down, what was 'masculine' became 'feminine' and vice versa. Such changes may be considered progress, but actually, they retain the gender binary; it is only reversed. Bending, shifting, dislocating or even only adjusting gender binaries appear to be precarious strategies.

Certainly, the book is not devoid of gender binaries. We do on occasion juxtapose men and women, even if we try to add quickly some modality: young, old, rural, modern, etc. However, a lot of effort has been put into destabilising these binaries, by various means. We have sought to diversify representations of gender, above all by showing how dichotomies are misleading and unstable signifiers of real-life practice. We have sought to subvert and dissolve gender-related dichotomies often deployed in relation to ICT, such as tool-toy and utility-play. We have also emphasised how exclusion may be produced through the way people 'do' their gender. The analysis in Chapter 3 of the community-focused inclusion initiative in Ardmore showed how the use of computers might be considered un-masculine.

Should we forget, dissolve, undo, de-gender or do away with gender, just stop writing about it? Of course, 'forgetting about gender' is in a way what we want most of all. In that case, gender would no longer be an important factor in life, no longer something to consider when designing products, hiring employees or writing about ICTs. On the other hand, how can we forget about gender when systematic inequalities between men and women remain in place, and when it is difficult to see practices as nothing but a result of formatting from gender-centred discourses? This seems an impossible situation. Whatever choice one makes, there seems to be a price to pay.

Or is there a way out? We need to discuss the catch between forgetting gender and overlooking gender-related injustices, on the one hand, and focusing on gender and gender-related injustices in such a way that an unwanted gender formatting is reproduced, on the other. Can we find ways to think about, design and develop initiatives which do not promote a liberalist 'we are all individuals' perception of gender (see Chapter 1), and which are sensitive to unfair gender practices, while at the same time avoid repeating and reinforcing existing gender dichotomies and stereotypes?

We claim to have made an effort to achieve this in this book. We do not know if our strategies of representation, emphasising diversity and complexity, are sufficient to avoid reproducing what we do not want to reproduce. In the end, you our readers must judge for yourselves. We hope you will share in the project of producing and nurturing more diverse and less gendered ICT-centred assemblages.

Brief overview of the SIGIS focused studies

Faulkner, Wendy and Tine Kleif. Edinburgh women's training course: An old idea still working.

The Edinburgh Women's Training Course (EWTC) was a woman-only IT-training programme, which sought to enable unemployed women living in disadvantaged circumstance to enter the labour market or further or higher education. The inclusion strategy involved women-only training, an approach that has had strong advocates in the past but has fallen from fashion to a considerable degree. Its users are disadvantaged women, including groups who may also be amongst those with least access to ICTs and ICT skills. Its approach was holistic in that the organisation and content of the course met many of the wider needs of its users, such as providing childcare and building self-esteem, alongside their digital inclusion need. The course is deemed a success in so much as the majority of its graduates find employment or go on to further or higher education. Since the women were unemployable and do not generally believe in themselves when they come on the course, these are very significant achievements.

Faulkner, Wendy and Tine Kleif. Included women, excluded men: Users and non-users of rural community resource centres.

The Ardmore Network established public access to computer technology with on-site support workers in remote rural communities. It offered an interesting example of a 'public sector-de facto' gender inclusion strategy since it was meant to be for 'everybody' but in practice more women than men made use of it. Also, the strategy addressed community development and, thus, a more political aspect of social inclusion, where most of the other cases address economic (labour market) inclusion. Moreover, the study also provided an interesting point of comparison since it highlights remote rural experiences of exclusion. The Network and its Resource Centres have been marketed to individuals and community groups. The range of actual users and the patterns of usage varied somewhat across the different localities – reflecting in part pre-existing differences between the localities. In all of the Centres studied, however, the majority of individual users were women.

Faulkner, Wendy and James Stewart. National ICT training for teachers: A case of 'gender blindness'

This was a non-gender differentiated, public sector strategy, which aimed to increase the effective use and integration of ICT in all aspects of primary and secondary education by training all teachers and librarians. It was more inclusive of women than men for the simple reason that women outnumber men in teaching, especially at primary level. This particular strategy was part of a sustained policy intervention by the UK Government, which aimed "to raise the standard of pupils' achievements by raising the expertise of serving teachers in the use of ICT in subject teaching to the level expected of newly qualified teachers". Virtually all teachers not recently qualified or already expert in ICT followed the training at some point over a period from 2000 to end 2002. In practice, the training actually delivered was relatively low level and undemanding; it appears to have been most valued by those who had little or no prior competence and/or confidence in using computers. To this degree, the strategy fell well short of meeting the rather sophisticated expected outcomes set for it. However, in the context of the wider policy initiatives, the strategy has no doubt contributed to increasing and generalising the use of ICT considerably in a range of activities, and to foregrounding the potential role of technology in education.

Fortunati, Leopoldina. 'The new cinderellas': How public administration has been attempting to include women as well

The object of this study was to analyse which inclusion strategies for women were produced in certain crucial areas of public administration. Public administration is particularly important because it has a particular responsibility for inclusion of women (and all citizens) in the Information Society. The analysis was conducted by focusing on four significant levels in the bureaucratic system: a) a national event, COMPA, the Public Communications Exhibition that is held every year in Bologna; b) The Bologna city council network, Iperbole, one of the most important networks at a European level; c) the Treviso Prefecture, a provincial official body; and d) the site of a small town council, Montecchio Maggiore (Vr).

Fortunati, Leopoldina. Lifelong education and strategies of gender inclusion in the Information Society

Central elements in strategies of women's inclusion in the Information Society are, on the one hand, teaching to use the computer and the Internet, and on the other hand, women that are learning about them. Three emblematic examples of the same process were examined, each of a different size: 1) Mediamente, the most important TV site and transmission of RAI Educational, which has informed surfers and TV viewers about the new technologies and illustrated their possible aims, modes and uses; 2) The Bologna Women's Bookshop, responsible for a series of concrete initiatives, among which Server Donne, an Internet Tearoom exclusively for women, and computer beginners' programmes aimed at this kind of user; 3) the organisation of a computer beginners' courses for parents in secondary schools in the Verona area. Thus, the study analysed three different strategies of women's inclusion put into practice at three distinct points of the lifelong educational process, to see how effective the inclusive strategies were, and at the same time to reconstruct the complexity of the problems involved.

Fortunati, Leopoldina. The Information Society at the service of tradition

This study concerns Donna Moderna, the most widely read women's weekly in Italy, and analyses whether and how it acted (via its website) as a driving force in facilitating women's access to and use of ICT. 'Donna Moderna' created one of the first and most active Italian women's magazine on-line sites. The study examines whether the Donna Moderna web-site: a) served to increase the interest of its women readers to learn about and use the computer and the Internet; b) if yes, whether the interest in ICT was of a general kind or whether it was articulated according to the specific characteristics of the readers; c) whether Donna Moderna pursued a policy of information about the ICT and the response from its readership in this regard; d) whether it put any experts at the disposal of readers/audiences to help and promote this approach to ICT; and e) whether it received specific suggestions, feedback or had been the object of pressure on the part of readers to change or strengthen its digital development.

Fortunati, Leopoldina. Women and the new economy: The Work Up case study
This study analyses Work Up, a firm that was created and run by a woman, Patrizia Bizzotto, an electronics engineer, who employed almost exclusively women. This firm offered various Internet services, web project consultancy, e-commerce and promotion services. It was a 'high-tech' firm, which contributed directly to the making of the Information Society. This company was examined as a potential new strategy of women's inclusion in the difficult world of labour in the new economy. The strong feminine character of this firm has produced innovative experiments. In addition, by focusing on women in the process of training, it helped in formulating digital services specifically directed at women.

Fortunati, Leopoldina. Women and the media system in society, a case study: The Donna Moderna site forums
This study investigates the inclusion strategy of the women's weekly magazine Donna Moderna. The discussion forums on the site were studied, to analyse the competence of users of this site in using the Internet. The strategy turned out to be very effective if we look at the high number of women who used this site. It must, however, be stressed that they were younger women, rather than adults, with a high level of education. Competence in using the Internet and computers turns out to be quite high, seeing the great ease with which the women used the forum and the related ICT technologies.

Fortunati, Leopoldina and Roberto A. de Luca. Behind the mask
This study investigates the kind of women's inclusion strategies offered by a voluntary sector website, www.Lupus.it. The site provided information about a rare disease that affects mainly women, especially young girls. Most doctors know little about this disease, so the Internet has become a vital place to find information about the disease and share stories and

experiences. Here women can find the latest knowledge and very useful advice about how to deal with it. We analysed whether contents relevant to women interests, in this case health, manage to help (especially young) women to overcome the diffidence and fear that much research has shown as quite common responses by women towards the Internet.

Fortunati, Leopoldina. The 'Wolverines' discuss: The Lupus user's case study

This study explores the effects of the website www.lupus.it from a user perspective. The site is aimed at women who have the rare disease 'lupus', which strikes only women. Although the website was not created for the purpose of digital inclusion, it did work out that way. The sole fact that the founders managed to build a virtual space aimed mainly at a specifically feminine area of interest permitted women who have some knowledge of the computer to have a virtual space for themselves. In this way the women using the site were able to create their own virtual community. The strategy of self-inclusion in the Information Society by the women users of the site greatly raised their level of competence in the use of the computer and Internet.

Fortunati, Leopoldina and Anna Maria Manganelli. From the GSM to the UNMTS: Is it a path towards women?

This study deals with the strategy of inclusion of women that operators were producing in the transition from the GSM to the UMTS. We interviewed the research team that was managing the transition to the next generation of mobiles inside the Telecom Italia Lab. The aim of this research group was to incorporate the point of view, desires and fantasies of users, right from the first moments of production of technological devices. The study examines how new proposals for services for the UMTS seemed to be placing women at the centre of the technological project even if only partially and with some ambiguity. The invention of the OSA platform and local portals, together with recognition of the growing spending power and power of determination of women, were all elements that should have led to the creation of specific terminals and services dedicated to women themselves.

Fortunati, Leopoldina and Anna Maria Manganelli. Women in the digitalisation of public administration

This study analyses experiences of a cross-section of women working in public administration. One of the main findings in this case study is that the strategies of women's inclusion in the public administration oscillate between being de facto strategies and being strategies of self-inclusion. The unsatisfactory and non-standardised way women learned to use computers and the Internet is to be attributed to the disorderly way that computers spread through the social body, going either through self-teaching or through ad hoc teachers (husbands, friends, colleagues) who are not very competent themselves. If this problem is not resolved, we risk that these new technologies are not used to their full potential.

Gansmo, Helen Jøsok. Girls and computing as national educational strategy in Norway

This study analyses the Norwegian Ministry of Education's strategy to achieve gradual inclusion of all primary and secondary school pupils, especially girls, to become competent computer users. This meant achieving a 50/50 distribution of boys and girls in front of the computer and avoiding new digital divides due to gender, which again meant ensuring everyone's access. Further, everyone was expected to be included through integrating computing in all school subjects. To secure the inclusion of girls, special girl groups, and initiatives where the girls were taught ICT first, were suggested. The initiatives were taken by the Ministry of Education, while the responsibility was placed on the individual schools and the local governments.

Gansmo, Helen Jøsok. Limits of state feminism: Chaotic translations of the 'Girls and computing' problem

This study analyses the way Norwegian schools reacted to the strategy of the Ministry of Education (see above study) to get girls included as users of computers/ICT. Thus, it is an exploration of the extent to which the schools themselves developed such inclusion strategies and, if so, how they were shaped. The study was based on interviews with managers at seven schools. Despite large differences among these schools, three main stories were detected. The first group of schools turned the instructions from the Ministry down due to lack of financial support for implementation of the instructions. The second group of schools worked with what they had and was actively seeking more funding or project participation. The last group is labelled 'The innovative schools' due to their innovative, but 'disobedient' implementation strategy.

Gansmo, Helen Jøsok. Computing: excludingly boring at school – includingly cool at home

This study looks at how pupils in Norwegian secondary schools used and thought about computers and computer users in their school and everyday lives. The main strategy of Norwegian government has been to secure access to school computers for everyone, especially girls, and to implement computing in the teaching of the different subjects. Schools managed to make computing in school transgress gender and becoming trivial (as in routine), but this was ironically done in a very boring manner, which in itself can be excluding for everybody. At home, pupils used chat and downloaded music from the Internet – rather demanding tasks which require skills and lack of technophobia – with great enthusiasm. These tasks were constructed as gender neutral (non-)computing activities. Thus, the pupils were included rather through leisure computing than school computing.

Gansmo, Helen Jøsok, Hege Nordli and Knut H. Sørensen. The gender game: A study of Norwegian computer game designers
This study examines the design of computer games to analyse whether and how gendered assumptions work as design criteria, and whether the games industry was making efforts to provide products attractive to girls and women. The study is based on interviews in four Norwegian computer game companies. It is commonly believed that computer games are one of the most important gateways to become a skilled and enthusiastic user of computers. In this study, we try to contribute to an improved understanding of the gendered dynamics of computer games design, with an emphasis on what the industry considered appropriate strategies of inclusion towards potential women users.

Hestflått, Kristin. Strategies of inclusion in three web-based magazines: For 'women', for 'the advanced' and for 'the home user'
This study focuses on how producers of web magazines use different strategies of inclusion to recruit readers and make them users of the Internet. Mainly, we analysed inclusion strategies used by a women's magazine – *Femme* – that primarily was web-based. To explore eventual particularities of the inclusion strategies employed by this magazine, we have done a comparison with inclusion strategies applied by two computer magazines also with websites. *Femme* had a target group of "women that are highly educated, innovative and conscious of quality". The computer magazine *HjemmePC* aimed at "ordinary users", while the second computer magazine *PCPro* tried to reach a more knowledgeable, professional segment. In this way, we compare one approach that is consciously developed to reach women readers (inclusion of women in particular) with approaches that cater either for a general audience (everyone, including women) or for a specialist, presumably an audience

of predominantly mwn (but women are of course allowed too). What we found was that *Femme's* strategies recruited and socialised women to interact and communicate on the Internet. However, through the websites of the computer magazines, men were invited to a much wider culture of ICT.

Hynes, Deirdre. The role of computer courses in the domestication of the computer

This study investigates a computer course that was offered to all parents of children in schools. The majority of those who availed themselves of the courses were in fact women, probably because the courses were run during the school term, during the school day and the availability of childcare was strongly advertised. The course facilitated the domestication of the technology by teaching the women relevant skills. The women stated that doing the course removed some of the barriers or obstacles facing them and opened up a conduit whereby a relationship with the technology could be constructed.

Kerr, Aphra. Girls just want to have fun!

Four inclusion strategies were explored: inclusion through console advertising, through console re-design, through social networks and through game design. The first inclusion story is that gender-neutral advertising for consoles largely was ineffectual in terms of recruiting women to become owners. The second strategy explored was about inclusion through console re-design. Interviews indicated that the size, shape and price of game consoles were important but that the range of games available on the platform also was an important factor motivating women's purchasing decision. The third inclusion story indicates that local social networks appear to play a crucial role in enrolling and socialising women players. In the final inclusion story, interviewees described that they were made aware of the dominance of men in this culture in game shops, through their own social networks, but especially through the representations offered to them by game designers, publishers and marketing professionals.

Kleif, Tine and Wendy Faulkner. Rural community resource centres: A case of 'de facto' women's inclusion in the Information Society

The Ardmore Network is a network of Community Resource Centres combining public access ICTs with on-site Support Workers offering one-to-one support and training in a remote rural community in Scotland. Although targeted on 'everybody', the strategy included more women than men. This can be explained firstly by the gender segregation of (paid) labour locally, which meant that many men did not need ICT skills for their jobs and often were more computer reticent than women of the same age and class. Secondly, because Ardmore Network had not provided appropriate role models or created social spaces which felt comfortable to computer reticent men. The contribution of the Ardmore Network to local ICT capability building was greater in relation to older age groups, for whom it played the role of providing local ICT expert and network.

Kleif, Tine and Wendy Faulkner. 'I am not the same person!': User study of a women-only training course

The Edinburgh Women's Training Course (EWTC) was a woman-only IT-training programme, which sought to enable unemployed women living in disadvantaged circumstances to enter the labour market or to get access to further or higher education. EWTC provided women-only training in IT skills as a route to labour market/educational inclusion and personal development for particular groups of socially and economically excluded women. The majority of its graduates found employment or went on to further or higher education. They were also empowered more generally in their lives through the efforts of the training programme to build their self-esteem and confidence.

Lagesen, Vivian A. Squares and circles: Getting women into computer science

This study concerns an initiative intended to increase the number of women students in a computer science and engineering programme at NTNU. The initiative consisted of a range of different actions. The plurality of actions and the different perceptions and understanding of the initiative among the actors were comprised in a strategy that achieved a substantial rise in the number of women students. Thus, it was regarded as a great success. However, the initiative may be viewed as theoretically and politically questionable, since it tended to naturalise and dichotomise gender in quite strict terms. This is shown through an analysis of the advertising campaigns that was made by the initiative.

Lagesen, Vivian A. Computer science – careers or computing? Inclusion through 'secularization' of ICT

The Women and Computing Initiative at NTNU aimed to get more women to study computer science through, amongst other things, advertiseing campaigns. The inclusion strategy of the initiative achieved a substantial rise in the number of women students and was regarded as a great success. Many of the women students who entered the programme reported that they had been influenced by the campaign when they chose to study computer science. The attractiveness of computer science was related to good career prospects and a relative high status of the education. Also, the initiative signalled that women were welcome in the programme, that there already were quite a lot of women students and that they did not need to have a lot previous knowledge. However, the content of the advertising material was characterised by an identification of the readers as not so interested in science and technology. It overlooked the fact that most of the women entering the programme were motivated to study science or engineering, and that these women would be more interested in factual information about the education, rather than criticism of hackers and their mentality.

MacKeogh, Carol. Women in technology and science role model project

This case study explores the concept of role modelling as a means to create positive work images of women in ICT for dissemination among secondary school girls. For a number of years, Women In Technology and Science in Ireland (WITS) have been supporting role model days in institutes of higher education and universities across the country. As well as looking at the underpinning concept, this study also explores the practicality of a Role Model Pack whereby third level institutions would be able to self-administer a role model day.

MacKeogh, Carol. The Cork Institute of Technology and Fastrack to IT: Initiatives for the lone parents and the long-term unemployed

This study looks at two separate initiatives to encourage women and the long-term unemployed to consider careers in information technology, and science and technology. One initiative was based in a particular institution (Cork Institute of Technology (CIT)) while the other was a certification process that was operated in a number of educational institutions (Fastrack to IT (FIT)). In both cases the aims behind the initiatives were primarily to increase numbers in the relevant institutions or organisations. While both initiatives argued a degree of social motivation, FIT was a response to the needs of industry due to a skills shortage, and the CIT initiative can be seen as a part of a recruitment drive for the third level institution.

MacKeogh, Carol. Community application of communication technologies (CAIT): The case of the Cork women's education initiative (CWEI)
This study looks at a community-based series of initiatives that were aimed at allowing 'late adopters' the opportunity to try out new technologies. In particular, it concentrated on a community group of women who were among those that gained funding to create computer courses. Cognisant of the potential impact of a digital divide on democracy, given the development of e-government, the Irish government created this initiative in order to help specific targeted groups to explore the relevance of technology to their lives.

MacKeogh, Carol: Women in technology international (WITI)
Women in Technology International (WITI) was established as a non-profit professional organisation for women working in the technology sector in the USA over a decade ago. Its aim is to provide these women with "inspiration, education, conferences, on-line services, publications and an exceptional world-wide network of resources". It claims to be the first and only international organisation solely dedicated to advancing women through technology, and that its network extends to 75,000 women. At the time of the study, WITI had 34 regional chapters in the US and this had been the extent of its 'international' reach. Later, branches had been set up in England, Australia and in March 2002, an Irish branch was established. This study examines the initial stages in the establishment of WITI in Ireland. It explores the approach that was adopted, the views of those that influenced this formative period and the needs that they identified in the sector. This voluntary initiative has had a very successful launch both in terms of numbers and enthusiasm but a sustained role for the organisation will depend on its relevance to Irish women working in the sector.

MacKeogh, Carol. iVenus.com: Creating a space for women on the web
iVenus is a women's on-line magazine that was set up by a private publishing company around 2000. It hosts a vibrant discussion board, which may be considered an initiative to create gender inclusion due to its specific relevant-to-women ethos. In a cyber world where the men 'hacker ethic' often dominates, this discussion board was developed by the users themselves in a way that they found relevant and supportive. Based on interviews with nine staff members, a number of issues arose that portray various aspects of the relationship between the staff and the users. The report examines the sources that staff used to develop their

profiles of their audience and queries what role the audience played in the choices made regarding the technology and management system that ran the site. It explores how the staff perceived their users both as readers of their magazine and as women using new technology. Staff views are placed in the broader context of both their work practices and their views on gender and the digital divide in general.

MacKeogh, Carol. Win win situation: Human resources perspective on gender in the ICT workforce

This study investigates the policies and initiatives of the Human Resources sections in six major ICT companies operating in Ireland. These companies were all multinationals and had previously been identified as sources of greater gender inclusion than indigenous companies. The study involves three separate strands. It set about trying to map the current situation with regard to gender streaming, to investigate the human resources ethos in terms of gender inclusion and to explore the nature and success of specific inclusion policies. In a climate of minimal unionization or other regulation, a key question was whether or how the initiatives tackled gender exclusion or whether inclusion polices formed part of a human resources discourse. In all six companies, the human resources departments reported that gender was no longer as central an issue as it had been in the past. Often this was said to be because it now was subsumed under the umbrella of 'diversity'. Conceptually, this appears to have entailed a shift away from gender auditing to a focus on greater uniformity of treatment. This may have suited prevailing concerns around gender essentialism in that women did not feel that they were a 'problem' or in need of special help to get on in their careers.

MacKeogh, Carol and Barbara O'Connor. 'Performing on the boards': Female users of an online discussion board

In this study the on-line discussion boards of a women's on-line magazine are explored. This study regards the discussion boards as a key impetus in introducing women to an interactive role on-line. The site appeared to provide an extremely open forum for the users to create their own space. It has allowed these particular women to create a 'female' space that they find welcoming and supportive. The web has traditionally been seen as a men's preserve so that the appropriation of this space by these women (given that the board is un-moderated) was an initiative in itself. There were, however, some constraints. The accessibility, as a website, was also the source of its vulnerability which, in turn, necessitated a high degree of self-regulation.

Nordli, Hege. The Gathering: Computer parties as means for gender inclusion

This study examines an annual computer party called 'The Gathering', the largest such party in Norway as well as Europe. 'The Gathering' was organised by a voluntary organisation that emphasised the importance of young people acquiring good computer skills. The participants averaged between 17 and 23 years old. Parties like this used to be an all men arena. It was still dominated by men, but the attendance by women had grown to around 10 %. KANDU, the organisers behind 'The Gathering', had an inclusion-for-all strategy. In addition, however, the organisers also made some specific inclusion-of-women efforts. One important inclusion aspect of 'The Gathering' was the way in which it presented computing as a social and fun activity.

Nordli, Hege. The Gathering experience: A user study of a computer party

The Gathering was a computer party for computer interested young people in Norway. Even though the computer was an important element, the participants most of all came there to meet up with friends and get new friends. The attendance of women was between 10 and 15% and it had not increased a lot the previous couple of years. Even though the (women) participants see the gender gap, they claimed that this would change by itself since more and more girls were introduced to the computer from an early age and they do not experience it as a problem. Being a serious computer enthusiast and a girl did not seem to be a problem to them. Yet the girls were very often seen by the boys as just being into chat and not doing serious computing. They were engaged in the 'wrong' type of activities and were therefore not given much credit.

Nordli, Hege and Knut H. Sørensen. Diffusion as inclusion? How adult men and women become users of mobile phones

Although the mobile telephone was initially an artefact used mainly for business purposes and mainly by men, women became included as users on a large scale and over a quite brief period. In this study we found that there has been a gendered practice when looking at recruitment into the constituency of mobile phone users. Men tended more often to have been earlier adopters and also to be the ones that had the better model and a mobile paid by their employer, whereas a large number of women interviewees had received their mobile phones as a gift or as a second-hand transfer. However, it was not perceived as gender-inauthentic for women to own or use mobiles, and women seemed to become just as easily socialised into the community of mobile users as men did. We also observed a gender difference in the way men and women accounted for their use of a mobile and for their skills in using the mobiles. These differences were not based on the actual practices of the men and women interviewees.

Pitt, Lisa. IT Beat: Bringing pop and glam to IT
This study concerns an innovative inclusion strategy, 'IT Beat', aimed at changing young girls' perception of IT and building their computer skills in the UK. The idea of trying to improve IT's poor image came in response to a drop in the number of women working in the sector. Special consideration was given to the key motivations behind the initiative as well as its particular features. The strategy of the 'IT Beat' initiative itself was split into a number of phases which under the guise of glamour and fun hoped to change young girls' perception of IT, and significantly too their outlook on careers in the field. It included Competitions where winners were invited to an overnight event at the Science Museum in London, and opportunities for girls to meet pop stars and other women celebrities involved in a broad range of professions, in which technology was an important aspect. The event was also promoted as the world's biggest slumber party.

Pitt, Lisa. New media, old world: The Untold story
This study examines Untold, a UK-based group offering women digital designers a platform to showcase their work and share experiences of working in the field. The study highlights the persistence of old structures of gender inequality in the new media industries, and how Untold responded to this. The group's work underlined processes and practices that appeared to present particular disadvantages to women, including limited exposure, restricted job opportunities or gender insensitive working environments. Untold aimed to address such concerns by promoting talented women digital designers in the form of an on-line exhibition, by engaging individuals at all levels through public meetings and discussions, and by raising awareness of the situation. The Untold strategy was therefore multifaceted in both its content and aims. The findings also revealed Untold's delicate balancing of feminist, post-feminist and post-modern discourses in its effort to present itself as an inclusion, rather than exclusion, strategy.

Pitt, Lisa. Don't leave IT to the boys!

The IT Beat user study follows on from the IT Beat strategy study. IT Beat was an initiative put together by e-skills, the agency responsible for ensuring the skills levels of the UK workforce in IT and telecommunications, with the support of the UK Government, the (London-based) Science Museum and a number of private companies, including IBM. The initiative aimed to change young girls' perception of IT and jobs in the IT sector, by inviting girls aged 11–15 to take part in a competition, where they were asked to design a website for their favourite pop star. The value of the strategy was to highlight girls' enthusiasm for technology and their ability to use technology creatively and to enjoy using it.

Pitt, Lisa. Give and take: A user perspective of WITI

The study provides an insight into the user experience in the UK of an organisation aimed at retaining and supporting women in technology (WITI), rather than working towards inclusion at an entry level. WITI provided particular skills, training and confidence, yet some of the women interviewed in this study appeared to be very confident. It is difficult to tell whether this confidence has come from their participation in WITI or whether it is a characteristic necessary for reaching their level in a male-dominated environment. The WITI members interviewed, even if aware of the gender imbalance in their sector, were keen not to make WITI a gender issue. WITI was portrayed rather as emphasising choice and as providing women with the support, confidence and skills required to succeed in a technology and/or business environment. For some, moving beyond gender and making it irrelevant was seen as an ideal situation, whilst for others highlighting the value of feminine attributes in the workplace, such as cooperative working, needed greater attention.

Pitt, Lisa. Being in digital design: The Untold experience

Untold was an initiative initially devised to support women digital designers through an exhibition and had developed into a platform for discussion on gender and technology, a social and professional network and a community of individuals working in and around digital design. Untold was targeted at women as they have been found to suffer from inequalities related predominantly to the gender imbalance in the sector. Untold however did not exclude men. The strategy was perceived very positively by those involved, primarily because they felt that inappropriate behaviour and the gender imbalance in the sector in general needed particular and urgent attention.

Rommes, Els. 'I don't know how to fit it into my life': Courses, computers and the Internet for 'everybody' in Amsterdam

In this study, a project by several Amsterdam-based organisations to 'diminish the digital divide' by offering free introductory courses computers and the Internet, is analysed. It was a typical example of a large-scale national project by the Dutch Government to support local initiatives to provide access to public terminals and to offer courses to people who do not have easy access to computers and the Internet. The project consisted mainly of free introductory courses 'Computers and the Internet' and of providing more public access terminals throughout the area in which the project took place. The project was successful in reaching people with a diverse background, e.g. a high percentage of ethnic minorities, elderly people. About 70% of the course participants were women.

Rommes, Els. The Women's Square in a digital city: 'Inclusion for all' versus 'inclusion for women'

The Women's Square was a location on the Internet where information from and for a wide range of women's organisations in the Netherlands was gathered. Next to this, the site offered the mailing list 'womenmail', that initially was only accessible to women, the moderated discussion group femail.mod and a 'prikbord', a place where women could leave announcements about women's activities. They had an 'inclusion for women' strategy. The target group of the Women's Square was mainly women active in the women's movement. Because the Women's Square was part of the interface of Amsterdam's digital city 'De Digitale Stad' (DDS) and had to fit into this interface, the founders of the Women's Square negotiated and cooperated with the designers of DDS.

Rommes, Els. 'I don't know how to fit it into my life': The gap between the inclusion initiative 'Introduction courses computers and the internet' and the personal stories of the excluded

This is a study of a free course to learn the basics of computers and the Internet. In general, all the informants were satisfied with the course. For some informants, the consequence of the course was that they decided that computers and the Internet did not fit their lives; they became informed rejecters. If these rejecters needed to be included, it might be a point of consideration to have aimed the course more at exploring what individual participants could use the computers and the Internet for, before attempts were even made to teach participants the basic skills. Discovering if and how the computer fits the every day lives of individuals might have helped more than any kind of skill training. It seemed that the largest influence of the course 'getting to know computers and the Internet' was that the high expectations some of the informants had about the Internet were somewhat diminished.

Rommes, Els, Nelly Oudshoorn and Marcelle Stienstra. KidCom designer case
The goal of the KidCom project (launched by Philips) was defined as "finding out what attracts girls aged 7 to 12 and use these findings to develop a personal electronic device". In this project, the composition of the design-team, the design methodology and the user representations were examined in detail. Children influenced the development trajectory of the projects and the type of product developed. Gender was treated in an essentialist way, as the designers spent much of their effort trying to define in which ways girls 'are' different from boys. Although the KidCom was developed for the electronic games market, the designers presented the KidCom not as a game but as a communication device – as the name KidCom suggests.

Rommes, Els and Irma van Slooten. Women on the Web user study
The objective of Women on the Web (WOW) was to make computers and the Internet more accessible for women by facilitating the provision of information to women and through socialisation activities with mailing lists. The women-only social community that WOW facilitated, was appreciated by the respondents, as they felt lonely in a men's world, they felt like men were degrading them and their level of self-confidence and their belief in the power of women was generally low before they entered the WOW community. WOW was a successful inclusion strategy, as almost all respondents, including starters and ICT professionals, used computers and the Internet more, had more fun and used more different facilities because of WOW. Moreover, more than half of the respondents gained self-confidence regarding computers and the Internet, and felt more independent and empowered because of WOW and the women role-models that it provided.

van Slooten, Irma. Vrouwen Vak Scholen designer case: Women's vocational training in the Netherlands
This study explores the design choices behind the courses of Vrouwen Vak Scholen (VVS) the Netherlands, ICT courses in particular. VVS the Netherlands are Dutch vocational centres for women. Training at VVS was given to conform with a reasoned formula. Key elements were labour

market orientation, flexibility in entrance criteria, short training, student counselling and women-only training. These key elements together formed a solid base for intense 1 year courses. In these courses, women were trained to become competent professionals who can stand up to other candidates on the labour market. It looks as though VVS succeeded in doing that. According to their numbers at the time of the study, 80% of the graduates found regular employment within one year. However, due to the recession and changing government policy, VVS faced problems. Because of this VVS had to compromise, which negatively affected the original strategy.

van Slooten, Irma. The Libelle website: A case study on the website of the Dutch women's magazine Libelle
This is a study of the 'Libelle' website, linked to the widely read Dutch women's magazine 'Libelle'. This magazine is targeted at middle-class, heterosexual women who are 35–55 years old, middle-high educated, working part-time and having young children at home. Libelle can be considered as a quite traditional women's magazine, but its website offered a lot of high quality information on computers and the Internet. The study of Libelle provides an opportunity to analyse some novel strategies for including non-users into the Information Society.

van Slooten, Irma. Vrouwen Vak Scholen: User study of women's vocational training in the Netherlands
In this study the experiences of (graduated) students of ICT courses of a Dutch women's vocational training centre were analysed. Most of the women were satisfied with the ICT courses offered at VVS for four reasons. Firstly, women pointed to the technical, practical and social content of the courses. Secondly, they mentioned that, in order to facilitate the combination of study and care tasks, VVS offered part-time courses, following term time and school hours of the children. Thirdly, the Dutch Government made it financially possible for these women to attend the courses. Finally, the interviewees indicated the relevance of the 'women-only' strategy for learning about technical things. Statistics indicated that 80% of the women who followed an ICT course at VVS found a job in the ICT field.

van Slooten, Irma and Els Rommes. Women on the web: Designer case
'Women on the Web-The Netherlands' (WOW) is a network for women on the Internet. The key objective of WOW has been to make computers and Internet more accessible to women. WOW tried to reach this goal

by providing mailing lists, courses, and information about computers and the Internet on their website. An important part of WOW was a variety of mailing lists. The subjects of these mailing lists varied from computer and computer-related subjects, to lists for enterprising women and women working as journalists. At the time of the study, 4424 women had subscribed to WOW. WOW tried to reach their objective by a range of inclusion activities, which comprised recruitment as well as socialisation instruments.

Stewart, James. Boys and girls stay in to play: Creating inclusive and exclusive computer entertainment for children

This study examines a small firm of interaction designers and artists who created and sold creative interactive computer-based products to publishers and other intermediaries in the market. The firm found its 'multilayer' approach to design and its particular graphic style attractive to both girls and boys, but this was not recognised by its intermediate customers. Despite producing some games, the firm did not sell or design for the mainstream gaming market, which it viewed as dominated by a technology-led paradigm and products for boys.

References to the SIGIS focused studies

The studies are published in the following three volumes:

Lie, M. and K. H. Sørensen (eds.) *Strategies of Inclusion: Gender in the Information Society Vol. I Experiences from public sector initiatives*, Report 2003-63, Trondheim, Norway: NTNU, Centre for Technology and Society (ISSN 0802-3581).

MacKeogh, C. and P. Preston (eds.) *Strategies of Inclusion: Gender in the Information Society Vol. II Experiences from private and voluntary sector initiatives*, Report 2003-65, Trondheim, Norway: NTNU, Centre for Technology and Society (ISSN 0802-3581).

Oudshoorn, N., E. Rommes and I. van Slooten (eds.) *Strategies of Inclusion: Gender in the Information Society Vol. III Surveys of Women's User Experience*, Report 2004-66, Trondheim, Norway: NTNU, Centre for Technology and Society (ISSN 0802-3581).

These are the reports from the SIGIS studies, which are referred to by names and year in square brackets, e.g. [Faulkner and Kleif 2003]:

Faulkner, W. and T. Kleif. 2003. Edinburgh women's training course: An old idea still working, vol. 1, 201–236.

Faulkner, W. and T. Kleif. 2004. Included women, excluded men: Users and nonusers of rural community resource centres, vol. 3, 137–166.

Faulkner, W. and J. Stewart. 2003. National ICT training for teachers: A case of 'gender blindness', vol. 1, 273–298.

Fortunati, L. 2003a. 'The new cinderellas': How public administration has been attempting to include women as well, vol. 1, 327–358.

Fortunati, L. 2003b. Lifelong education and strategies of gender inclusion in the information society, vol. 1, 359–400.

Fortunati, L. 2003c. The information society at the service of tradition, vol. 2, 283–302.
Fortunati, L. 2003d. Women and the new economy. The work up case study, vol. 2, 303–328.
Fortunati, L. 2004a. Women and the media system in society. A case study: The Donna Moderna site forums, vol. 3, 233–254.
Fortunati, L. 2004b. The "Wolwerines" discuss. The Lupus user's case study, vol. 3, 244–292.
Fortunati, L. and R. A. de Luca. 2003. Behind the mask, vol. 1, 299–326.
Fortunati, L. and A. M. Manganelli. 2003. From the GSM to the UNMTS: Is it a path towards women?, vol. 2, 329–360.
Fortunati, L. and A. M. Manganelli. 2004. Women in the digitalisation of public administration, vol. 3, 293–318.
Gansmo, H. J. 2003a. Girls and computing as national educational strategy in Norway, vol. 1, 59–86.
Gansmo, H. J. 2003b. Limits of state feminism: Chaotic translations of the "Girls and computing" problem, vol. 1, 87–118.
Gansmo, H. J. 2004. Computing: Excludingly boring at school—Includingly cool at home, vol. 3, 337–362.
Gansmo, H. J., H. Nordli and K. H. Sørensen. 2003. The gender game. A study of Norwegian computer game designers, vol. 2, 115–140.
Hestflått, K. 2003. Strategies of inclusion in three web-based magazines: For "women", for "the advanced" and for "the home user", vol. 2, 91–114.
Hynes, D. 2004. The role of computer courses in the domestication of the computer, vol. 3, 167–184.
Kerr, A. 2004. Girls just want to have fun!, vol. 3, 211–232.
Kleif, T. and W. Faulkner. 2003. Rural community resource centres: A case of "de facto" women's inclusion in the information society, vol. 1, 237–272.
Kleif, T. and W. Faulkner. 2004. "I am not the same person!." User study of a women-only training course, vol. 3, 107–136.
Lagesen, V. A. 2003. Squares and circles: Getting women into computer science, vol. 1, 119–150.
Lagesen, V. A. 2004. Computer science—Careers or computing?. Inclusion through "secularization" of ICT, vol. 3, 363–414.
MacKeogh, C. 2003a. Women in technology and science role model project, vol. 1, 401–418.
MacKeogh, C. 2003b. The Cork institute of technology and Fastrack to IT. Initiatives for the lone parents and the longterm unemployed, vol. 1, 419–438.
MacKeogh, C. 2003c. Community application of communication technologies (CAIT): The case of the Cork women's education initiative (CWEI), vol. 1, 439–456.
MacKeogh C. 2003d. Women in technology international (WITI), vol. 2, 361–384.
MacKeogh, C. 2003e. eVenos.com: Creating a space for women on the web, vol. 2, 385–404.
MacKeogh, C. 2003f. Win win situation. Human resources perspective on gender in the ICT workforce, vol. 2, 405–424.
MacKeogh, C. and B. O'Connor. 2004. "Performing on the boards": Female users of an online discussion board, vol. 3, 185–210.

Nordli, H. 2003. The Gathering: Computer parties as means for gender inclusion, vol. 2, 71–90.
Nordli, H. 2004. The Gathering experience: A user study of a computer party, vol. 3, 319–336.
Nordli, H. and K. H. Sørensen. 2004. Diffusion as inclusion? How adult men and women become users of mobile phones, vol. 3, 415–438.
Pitt, L. 2003a. IT beat: Bringing pop and glam to IT, vol. 2, 219–240.
Pitt, L. 2003b. New media, old world: The Untold story, vol. 2, 241–258.
Pitt, L. 2004a. Don't leave IT to the boys!, vol. 3, 65–76.
Pitt, L. 2004b. Give and take: A user perspective of WITI, vol. 3, 77–90.
Pitt, L. 2004c. Being in digital design: The Untold experience, vol. 3, 91–106.
Rommes, Els. 2003a. "I don't know how to fit it into my life". Courses, computers and the internet for "everybody" in Amsterdam, vol. 1, 151–166.
Rommes, E. 2003b. The women's square in a digital city: "Inclusion for all" versus "inclusion for women," vol. 1, 181–200.
Rommes, E. 2004. "I don't know how to fit it into my life". The gap between the inclusion initiative "Introduction courses computers and the internet" and the personal stories of the excluded, vol. 3, 491–510.
Rommes, E., N. Oudshoorn and M. Stienstra. 2003. KidCom designer case, vol. 2, 189–219.
Rommes, E. and I. van Slooten. 2004. Women on the web user case study, vol. 3, 459–490.
van Slooten, I. 2003a. Vrouwenvakscholen: Designer case. Women's vocational training in the Netherlands, vol. 1, 167–180.
van Slooten, I. 2003b. The Libelle website: A case study on the website of the Dutch women's magazine Libelle, vol. 2, 163–188.
van Slooten, I. 2004. Vrouwenvakscholen: User case women's vocational training in the Netherlands, vol. 3, 439–458.
van Slooten, I. and E. Rommes. 2003. Women on the web. Designer case, vol. 2, 141–162.
Stewart, J. 2003. Boys and girls stay in to play: Creating inclusive and exclusive computer entertainment for children, vol. 2, 259–282.

References

Adam, A. 1997. What should we do with cyberfeminism? In *Women in Computing*, ed. R. Lander and A. Adam, Exeter, UK: Intellect, pp. 17–27.

Ahuja, M. K. 2002. Women in the information technology profession: A literature review, synthesis and research agenda. *European Journal of Information Systems*, 11: 20–34.

Akrich, M. 1995. User representations: Practices, methods and sociology. In *Managing Technology in Society: The Approach of Constructive Technology Assessment*, ed. A. Rip et al. London, UK: Pinter Publishers, pp. 167–184.

Alaimo, S. and S. Hekman, eds. 2008. *Material Feminisms*, Bloomington, IN: Indiana University Press.

Amdam, A., M. Griffiths, C. Keogh, K. Moore, H. Richards and A. Tattersall. 2005. You don't have to be male to be able to work here, but it helps!' Gender and the IT labour market. In *The Gender Politics of ICT*, eds. J. Archibald, J. Emms, F. Grundy, J. Payne and E. Turner, London, Middlesex, UK: Middlesex University Press, pp. 283–296.

Archibald, J., J. Emms, F. Grundy, J. Pane and E. Turner, eds. 2005. *The Gender Politics of ICT*, London, Middlesex, UK: Middlesex University Press.

Aune, M. 1996. The computer in everyday life: Patterns of domestication of a new technology. In *Making Technology Our Own?: Domesticating Technology into Everyday Life*, ed. M. Lie and K. H. Sørensen. Oslo, Norway: Scandinavian University Press, pp. 91–120.

Bacchi, C. 1996. *The Politics of Affirmative Action: "Women," Equality & Category Politics*, Thousand Oaks, CA: Sage.

Bagilhole, B. 2006. Family-friendly policies and equal opportunities: A contradiction in terms?. *British Journal of Guidance and Counselling*, 34 (3): 327–343.

Balcita, A. M., D. L. Carver and M. L. Soffa. 2002. Shortchanging the future of information technology: The untapped resource. *SIGSCE Bulletin*, 34 (2): 32–36.

Balka, E. 2005. Tidying up loose ends: Theoretical and practical issues related to women's participation in technological design. In *User Involvement in Innovation Processes; Strategies and Limitations from a Socio-Technical Perspective*, ed. H. Rohracher, Munich, Germany: Profil Verlag, pp. 147–174.

Bakardjieva, M. 2005. *Internet Society: The Internet in Everyday Life*, Thousand Oaks, CA: Sage.

Bakardjieva, M. and R. Smith. 2001. The internet in everyday life: Computer networking from the standpoint of the domestic user. *New Media and Society*, 3: 67–83.

Barker, J. L. and W. Aspray. 2006. The state of research on girls and IT. In *Women and Information Technology: Research on Underrepresentation*, eds. J. M. Cohoon and W. Aspray, Cambridge, MA: The MIT Press, pp. 3–54.

Barriers to Equality in Academia: Women in Computer Science at MIT. 1983. Report prepared by female graduate students and research staff in the Laboratory for Computer Science and the Artificial Intelligence Laboratory at MIT.

Bartol, K. M. and W. Aspray. 2006. The transition of women from the academic world to the IT workplace: A review of the relevant research. In *Women and Information Technology: Research on Underrepresentation*, eds. J. M. Cohoon and W. Aspray, Cambridge, MA: The MIT Press, pp. 377–420.

Bech, U. 1997. *The Reinvention of Politics: Rethinking Modernity in the Global Social Order*, Cambridge, UK: Polity Press.

Bell, D. 1973. *The Coming of Post-industrial Society*, New.York: Basic Books.

Bell, D. and B. M. Kennedy, eds. 2000. *The Cybercultures Reader*, Abington, Cavendish Road, London, UK: Routledge.

Benschop, Y. and M. Verloo. 2006. Sisyphus' sisters: Can gender mainstreaming escape the genderedness of organizations?. *Journal of Gender Studies*, 15 (1): 19–33.

Berg, A. J. 1994. Technological flexibility: Bringing gender into technology (or was it the other way around). In *Bringing Technology Home: Gender and Technology in a Changing Europe*, eds. C. Cockburn and R. Fürst-Dilíc, Maidenhead, Berkshire, UK: Open University Press, pp. 94–110.

Berg, A. J. 1996. Digital Feminism. STS-report 25/1996. Trondheim, Norway: Centre for Technology and Society, NTNU.

Berg, M. 1998. The politics of technology: On bringing social theory into technological design. *Science, Technology, & Human Values*, 23 (4), 456–490.

Berg, V. A. L. 2000. Firkanter og rundinger. Kjønnskonstruksjoner blant kvinnelige dataingeniørstudenter ved NTNU, SKF-rapport 3/2000. Trondheim, Norway: Centre for Feminist and Gender Research, NTNU.

Berg, V. A. L. 2002. Squares and circles: Getting women into computer science, STS working paper 9/02. Trondheim, Norway: Centre for Technology and Society, NTNU.

Berg, V. A. L. and K. Kvaløy. 1998. A quantitative investigation of the well being and motivation among first year students at the department of computer and information science, together with an evaluation of the module "Know

Your Subject". Teknisk notat 8/98. Trondheim, Norway: Department of Computer Engineering and Information Science, NTNU.

Berker, T., M. Hartmann, Y. Punie and K. Ward, eds. 2006. *Domestication of Media and Technology*, Maidenhead, Berkshire, UK: Open University Press.

Berner, B. 1975. *Kvinnor innen teknik og naturvetenskap*. Lund, Sweden: Arkiv förlag.

Berner, B. 1996. *Sakarnas tilstånd: Kön, klass, teknisk ekspertis*, Stockholm, Sweden: Carlssson.

Beyer, S. 1990. Gender differences in the accuracy of self-evaluations of performance. *Journal of Personality and Social Psychology*, 59 (5): 960–970.

Beyer, S. 1999. Gender differences in the accuracy of grade expectancies and evaluations. *Sex Roles*, 41 (3-4): 279–296.

Beyer, S., K. Rynes and S. Haller. 2004. Deterrents to women taking computer science courses. *IEEE Technology and Society Magazine*, 23 (1): 21–28.

Bijker, W. E. and J. Law, eds. 1992. *Shaping technology/Building society*. Cambridge, MA: MIT Press.

Bilimoria, D., S. Joy and X. Liang. 2008. Breaking barriers and creating inclusiveness: Lessons of organizational transformation to advance women faculty in academic science and engineering. *Human Resource Management*, 47 (3): 423–441.

Blum, L. and C. Frieze. 2005. The evolving culture of computing: Similarity is the difference, *Frontiers*, 26 (1): 110–125.

Booth, C. and C. Bennett. 2002. Gender mainstreaming in the European Union. *European Journal of Women's studies*, 9 (4): 430–446.

Borge, M. A., A. Roth, G. T. Nichols and B. S. Nichols. 1980. Effects of gender, age, locus of control and self esteem on estimates of college grades. *Psychological Reports*, 47: 831–837.

Braidotti, R. 1994. *Nomadic Subjects: Embodiment and Sexual Difference in Contemporary Feminist Theory*. New York: Columbia University Press.

Braidotti, R. 2002. *Metamorphoses: Towards a Materialist Theory of Becoming*. Cambridge, UK: Polity.

Brants, K. and V. Frissen. 2005. Inclusion and exclusion in the information society. In *Media, Technology and Everyday Life in Europe: From Information to Communication*, ed. R. Silverstone. Aldershot, Hampshire, UK: Ashgate, pp. 21–32.

Bray, F. 2007. Gender and technology. *Annual Review of Anthropology*, 36: 37–53.

Bromseth, J. C. H. 2006. Genre trouble and the body that mattered. Negotiations of gender, sexuality and identity in a Scandinavian mailing list community for lesbian and bisexual women. Ph.D. thesis, STS-report 76/2006. Trondheim, Norway: Department of Interdisciplinary Studies of Culture, NTNU.

Brosnan, M. 1998. *Technophobia: The Psychological Impact of Information Technology*. London, UK: Routledge.

Brunner, C. and D. Bennett. 2002. The feminization of technology. In *Ghosts in the Machine: Women's Voices in Research with Technology*, eds. N. Yelland and A. Rubin. New York: Peter Lang, pp. 71–96.

Brunsdon, C. 1995. The role of soap opera in the development of feminist television scholarship. In *To Be Continued ... Soap Operas Around the World*, ed. R. Allen. London, UK: Routledge, pp. 49–65.

Buckingham, D. 2006. The electronic generation: Children and new media. In *The Handbook of New Media*, eds. L. A. Lievrouw and S. Livingstone. Thousand Oaks, CA: Sage, pp. 75–91.

Buckingham, D. and A. McFarlane. 2001. A Digitally Driven Curriculum? London, UK: Institute for Public Policy Research.

Butler, J. 1990. *Gender Trouble: Feminism and the Subversion of Identity.* London, UK: Routledge.

Callon, M., P. Lascoumes and Y. Barthe. 2009 [French original 2001]. *Acting in an Uncertain World: An Essay on Technical Democracy.* Cambridge, MA: MIT Press.

Carr, D. 2007. Contexts, pleasures, and preferences: Girls playing computer games. In *Worlds in Play: International Perspectives on Digital Games Research*, eds. S. D. Castell and J. Jenson. New York: Peter Lang Publishing, pp. 313–322.

Carter, R. and G. Kirkup. 1990. *Women in Engineering: A Good Place to Be?* London, UK: Macmillan.

Cassell, J. 1998. Storytelling as a nexus of change in the relationship between gender and technology: A feminist approach to software design. In *From Barbie to Mortal Kombat, Gender and Computer Games*, eds. J. Cassell and H. Jenkins. Cambridge, MA: MIT Press, pp. 298–327.

Cassell, J. and H. Jenkins. 1998a. Chess for girls?: Feminism and computer games. In *From Barbie to Mortal Kombat, Gender and Computer Games*, eds. J. Cassell and H. Jenkins. Cambridge, MA: MIT Press, pp. 2–45.

Cassell, J. and H. Jenkins, eds. 1998b. *From Barbie to Mortal Kombat. Gender and Computer Games.* Cambridge, MA: MIT Press.

Cassidy, M. F. 2001. Cyberspace meets domestic space: Personal computers, women's work, and the gendered territories of the family home. *Critical Studies in Media Communication*, 18 (1), 44–65.

Castell, S. D. and M. Bryson. 1998. Retooling play: Dystopia, dysphoria, and difference. In *From Barbie to Mortal Kombat, Gender and Computer Games*, eds. J. Cassell and H. Jenkins. Cambridge, MA: MIT Press, pp. 232–261.

Castells, M. 2001. *The Internet Galaxy: Reflections on the Internet, Business, and Society.* New York: Oxford University Press.

Castells, M. ed. 2004a. *The Network Society: A Cross-Cultural Perspective.* Cheltenham, UK: Edward Elgar.

Castells, M. 2004b. Why networks matter. In *Network Logic. Who governs in an interconnected world*, eds. H. McCarthy, P. Miller and P. Skidmore. London, UK: Demos, pp. 221–225.

Castells, M., M. Fernández-Ardèvol, J. L. Qiu and A. Sey. 2007. *Mobile Communication and Society: A Global Perspective.* Cambridge, MA: MIT Press.

Charles, M. and K. Bradley 2006. A matter of degree: Female underrepresentation in computer science programs cross-nationally. In *Women and Information Technology: Research on Underrepresentation*, eds. J. M. Cohoon and W. Aspray. Cambridge, MA: MIT Press, pp. 183–204.

Clegg, S. and D. Trayhurn. 1999. Gender and computing: Not the same old problem. *British Educational Research Journal*, 26 (1): 75–89.

Cockburn, C. 1983. *Brothers. Male Dominance and Technological Change.* London, UK: Pluto Press.

Cockburn, C. 1985. *Machinery of Dominance.* London, UK: Pluto Press.

Cockburn, C. 1989. Equal opportunities: The short and long agenda. *Industrial Relations Journal*, 20: 213–225.

Cockburn, C. and R. Fürst-Dilíc, eds. 1994. *Bringing Technology Home: Gender and Technology in a Changing Europe*. Maidenhead, Berkshire, UK: Open University Press.

Cockburn, C. and S. Ormrod. 1993. *Gender & Technology in the Making*. Thousand Oaks, CA: Sage.

Cohoon, J. M. 2002. Recruiting and retaining women in undergraduate computing majors. *SIGSCE Bulletin*, 34 (2): 48–53.

Cohoon, J. M. and W. Aspray. 2006a. A critical review of the research on women's participation in postsecondary computing education. In *Women and Information Technology: Research on Underrepresentation*, eds. J. M. Cohoon and W. Aspray. Cambridge, MA: MIT Press, pp. 137–180.

Cohoon, J. M. and W. Aspray, eds. 2006b. *Women and Information Technology: Research on Underrepresentation*. Cambridge, MA: MIT Press.

Corneliussen, H. Diskursens makt—Individets frihet. Kjønnede posisjoner i diskursen om data. (The power of the discourse—The freedom of the individual. Gendered positions in the discourse about computing). Ph.D. thesis. Bergen, Norway: Faculty of Humanities, University of Bergen.

Corneliussen, H. G. Forthcoming. *Gender-Technology Relations: Exploring Stability and Change*.Houndmills, Basingstoke, UK: Palgrave Macmillan.

Corneliussen, H. G. and J. W. Rettberg. eds. 2008. *Digital Culture, Play, and Identity: A World of Warcraft Reader*. Cambridge, MA: The MIT Press.

Correll, S. 1995. The ethnography of an electronic bar; the lesbian cafe. *Journal of Contemporary Ethnography*, 24 (3): 270–298.

Correll, S. J. 2004. Constraints into preferences: Gender, status, and emerging career aspirations. *American Sociological Review*, 69 (1): 93–113.

Countryman, J., A. Feldman, L. Kekelis and E. Spertus. 2002. Developing a hardware in programming curriculum for middle school girls. *SIGSCE Bulletin*, 34 (2): 44–48.

Crawford, S. 1983. The origin and development of a concept: The information society. *Bulletin of the Medical Library Association*, 71 (4): 380–385.

Creager, A. N. H., E. Lunbeck and L. Schiebinger. eds. 2001. *Feminism in Twentieth-Century Science, Technology, and Medicine*. Chicago, IL: The University of Chicago Press.

Cronin, C. and A. Roger. 1999. Theorizing progress: Women in science, engineering, and technology in higher education. *Journal of Research in Science Teaching*, 36 (6): 636–661.

Crutzen, C. K. M. 2007. Ambient intelligence, between heaven and hell. A transformative critical room? In *Gender Designs IT: Construction and Deconstruction of Information Society Technology*, eds. I. Zorn, S. Maass, E. Rommes, C. Schirmer and H. Schelhowe. Wiesbaden, Germany: VS Verlag für Sozialwissenschaften, pp. 65–78.

Dambrot, F. H., M. A. Watkins-Malek, S. M. Silling, R. S. Marshall, and J. A. Garver. 1985. Correlates of sex differences in attitudes toward and involvement with computers. *Journal of Vocational Behaviour*, 27: 71–86.

d'Haenens, L. and C. van Summeren. 2002. Ownership and use of "old" and new media among ethnic minority youth in the Netherlands: The role of the ethno-cultural position. *Communications*, 27: 365–393.

Dholakia, R. R. 2006. Gender and IT in the households: Evolving patterns of internet use in the United States. *The Information Society*, 22: 231–240.

Dholakia, R. R., N. Mundorf and N. Dholakia. 1996. Bringing infotainment home: Challenges and choices. In *New Infotainment Technologies in the Home, Demand-Side Perspectives*, eds. R. R. Dholakia, N. Mundorf and N. Dholakia. Mahwah, NJ: Lawrence Erlbaum, pp: 1–20.

Donath, J. and D. Boyd. 2004. Public displays of connection. *BT Technology Journal*, 22 (4): 71–82.

Donelan, H., C. Herman, K. Kear and G. Kirkup. 2009. Patterns of online networking for women's career development. *Gender in Management: An International Journal*, 24 (2): 92–111.

Drost, K. and M. Jorna. 2000. Empowering Women Through the Internet: Dutch Women Unite. Paper presented at the INET conference, Yokohama, Japan, July 18-21, 2000. Available: http://www.isoc.org/inet2000/cdproceedings/8a/8a_1.htm

Dryburgh, H. 2000. Underrepresentation of girls and women in computer science: Classification of 1990s research. *Educational Computer Research*, 23 (2): 181–202.

DTI (Department of Trade and Industry). 2004. Flexible Working in the IT Industry: Long Hours and Work-Life Balance at the Margins? Report to Department of Trade and Industry and the Women in IT Forum, carried out by Flexecutive.

Duerden Comeau, T. 2004. Cross National Comparison of Information Technology Employment, WANE Report No.5, September 2004. Available: http://www.wane.ca/PDF/IR5.pdf

Durndell, A., P. Glissov and G. Siann. 1995. Gender and computing—Persistent differences. *Educational Research*, 37 (3): 219–227.

Durndell, A., Z. Haag, D. Asenova and H. Laithwaite. 2000. Computer self efficacy and gender. In *Women, Work and Computerization: Charting a Course to the Future, IFIP Conference Proceedings*, eds. E. Balka and R. Smith. Boston, MA: Kluwer.

Edwards, P. 1990. The army and the microworld. *SIGNS: Journal of Women in Culture and Society*, 16 (1): 102–107.

Edwards, P. 1996. *The Closed World: Computers and the Politics of Discourse in Cold War America*. Cambridge, MA: MIT Press.

Ehn, P. 1988. *Work-Oriented Design of Computer Artifacts*. Stockholm, Sweden: Arbetslivscentrum.

Elkjær, B. 1992. Girls and information technology in Denmark—An account of a socially constructed problem. *Gender and Education*, 4 (1/2): 25–40.

Epstein, S. 1996. *Impure Science; Aids, Activism, and the Politics of Knowledge, Medicine and Society*. Berkeley, CA: University of California Press.

Etzkowitz, H., C. Kemelgor and B. Uzzi. 2000. *Athena Unbound: The Advancement of Women in Science and Technology*. Cambridge, UK: Cambridge University Press.

European Commission. 1998. Design for All and ICT Business Practice: Addressing the Barriers. Examples of Best Practice. Brussels: DG XIII-C/E Telematics Applications Programme "Design for All" for an Inclusive Information Society.

European Commission. 2006. She Figures 2006. Women and Science Statistics and Indicators. Brussels, Belgium: Directorate-General for Research.
Everett, A. 2004. On cyberfeminism and cyberwomanism: High-tech mediations of feminism's discontent. *SIGNS*, 30 (1): 1278–1286.
Fan, T.-S. and Li, Y.-C. 2005. Gender issues and computers: College computer science in education in Taiwan. *Computers and Education*, 44: 285–300.
Faulkner, W. 2000a. Dualisms, hierarchies and gender in engineering. *Social Studies of Science*, 30 (5): 759–792.
Faulkner, W. 2000b. The power and the pleasure? A research agenda for "making gender stick" to engineers. *Science, Technology, & Human Values*, 25 (1): 87–119.
Faulkner, W. 2001. The technology question in feminism: A view from feminist technology studies. *Women's Studies International Forum*, 24 (1): 79–95.
Faulkner, W. 2005. Belonging and becoming: Gendered processes in engineering. In *The Gender Politics of ICT*, eds. J. Archibald, J. Emms, F. Grundy, J. Payne and E. Turner. London, Middlesex, UK: Middlesex University Press.
Faulkner, W. 2007. "Nuts and bolts and people": Gender-troubled engineering identities. *Social Studies of Science*, 37 (3): 331–356.
Faulkner, W. 2009a. Doing gender in engineering workplace cultures: Part I—Observations from the field. *Engineering Studies*, 1 (1): 3–18.
Faulkner, W. 2009b. Doing gender in engineering workplace cultures: Part II—The in/visibility paradox. *Engineering Studies*, 1 (3): 169–189.
Faulkner, W. and T. Kleif. 2005. One size does not fit all! Digital in/exclusion in a rural community-based ICT initiative. *Journal of Adult and Community Learning*, 11 (1): 43–61.
Fenstermaker, S. and C. West, eds. 2002. *Doing Gender, Doing Difference. Inequality, Power, and Institutional Change*. London, UK: Routledge.
Foucault, M. 2007 [French original 2004, edited by M. Senellart]. *Security, Territory, Population. Lectures at the Collége de France 1977–1978*. Houndmills, Basingstoke, UK: Palgrave Macmillan.
Fox, M. F., D. G. Johnson and S. V. Rosser, eds. 2006. *Women, Gender and Technology*. Champaign, IL: University of Illinois Press.
Fox, M. F., G. Sonnert and I. Nikiforova. 2009. Successful programs for undergraduate women in science and engineering: Adapting versus adopting the institutional environment. *Research in Higher Education*, 50: 333–353.
Gabbert, P. and P. H. Meeker. 2002. Support communities for women in computing. *SIGSCE Bulletin*, 34 (2): 62–66.
Galpin, V., I. Sanders, H. Turner and B. Venter. 2003. Computer self-efficacy, gender and educational background in South Africa. *IEEE Technology and Society Magazine*, 22 (3): 43–48.
Gamba, M. and B. H. Kleiner. 2001. The old boys' network today. *International Journal of Sociology and Social Policy*, 21 (8): 101–107.
Gansmo, H. J. 1998. Det forvrengte dataspeilet. STS-rapport 36/1998. Trondheim, Norway: Centre for Technology and Society, NTNU.
Gansmo, H. J. 2004. Towards a happy ending for girls and computing? Ph.D. thesis, STS-report 67/2004. Trondheim, Norway: Centre for Technology and Society, NTNU.
Gansmo, H. J. 2006. Girls and computing. In *Encyclopedia of Gender and Infor-*

mation Technology, vol. I, ed. E. M. Trauth. Hershey, PA: Idea Group Reference, pp. 721–726.

Gansmo, H. J., V. A. Lagesen and K. H. Sørensen. 2003a. Forget the hacker? A critical re-appraisal of Norwegian studies of gender and ICT. In *He, She and IT Revisited: New Perspectives on Gender in the Information Society*, ed. M. Lie. Oslo, Norway: Gyldendal Akademisk, pp. 34–68.

Gansmo, H. J., V. A. Lagesen and K. H. Sørensen 2003b. Out of the boy's room? A critical analysis of the understanding of gender and ICT in Norway. *Nora*, 11 (3): 130–139.

Gibbs, S. 1998. Women on the web: A two-year journey in cyberspace. *The European Journal of Women's Studies*, 5: 253–261.

Giddens, A. 1976. *New Rules of Sociological Method: A Positive Critique of Interpretative Sociologies.* New York: Basic Books.

Gilchrist, A. 2004. *The Well-connected Community: A Networking Approach to Community Development.* Bristol, UK: Policy Press.

Gill, R. 2002. Cool, creative and egalitarian? Exploring gender in project based new media work in Europe. *Information, Communication and Society*, 5 (1): 71–89.

Gokhale, A. A. and K. Stier. 2004. Closing the gender gap in technical disciplines: An investigative study. *Journal of Women and Minorities in Science and Engineering*, 10, 149–150.

Granovetter, M. S. 1973. The strength of weak ties. *American Journal of Sociology*, 78 (May): 1360–1380.

Green, E. and L. Keeble. 2001. The technological story of a women's centre: A feminist model of user-centred design. In *Community Informatics: Shaping Computer-Mediated Social Relations*, eds. L. Keeble and B. D. Lauder. London, UK: Routledge, pp. 53–70.

Greenbaum, J. 1991. Toward participatory design: The head and the heart revisited. In *Women, Work and Computerization*, eds. I. V. Eriksson, B. A. Kitchenham and K. G. Tijdens. Amsterdam, the Netherlands: Elsevier, pp. 33–39.

Greene, A.-M. and G. Kirton. 2002. Advancing gender equality: The role of women-only trade union education. *Gender, Work and Organization*, 9 (1): 39–59.

Greenfield, S. 2002. SET FAIR: A Report on Women in Science, Engineering and Technology. London, UK: Department of Trade and Industry.

Gorriz, C. M. and C. Medina. 2000. Engaging girls with computers through software games. *Communications of the ACM*, 43 (1): 42–49.

Gressard, C. and B. Lloyd. 1987. An investigation of the effects of math anxiety and sex on computer attitudes. *School Science and Mathematics*, 87 (2): 125–135.

Grint, K. and R. Gill. eds. 1995. *The Gender-Technology Relation. Contemporary Theory and Research.* London, UK: Taylor & Francis.

Guardian, 2nd July 2003. Feminism: Outmoded and unpopular. Available: http://www.guardian.co.uk/uk_news/story/0,,989268,00.html

Gunkel, D. J. 2003. Second thoughts: Toward a critique of the digital divide. *New Media & Society*, 5 (4): 499–522.

Gurstein, M. 2003. Effective use: A community informatics strategy beyond the digital divide. *First Monday*, 8 (12): 1–27.

Hacker, S. 1989. *Pleasure, Power, and Technology: Some Tales of Gender, Engineering and the Cooperative Workplace.* Boston, MA: Unwin Hyman.

Hacker, S. 1990. *Doing It the Hard Way: Investigations of Gender and Technology*, eds. D. E. Smith and S. M. Turner. Boston. MA: Unwin Hyman.

Haddon, L. 1990. Researching gender and home computers. In *Technology and Everyday Life: Trajectories and Transformations*, Report No 5: Proceedings from a Workshop in Trondheim, Norway. Oslo, Norway: Norwegian Research Council for Science and the Humanities, pp. 89–108.

Haddon, L. and G. Paul. 2001. Design in the IT industry: The role of users. In *Technology and the Market: Demand, Users and Innovation*, eds. R. Coombs, K. Green, A. Richards and V. Walsh. Cheltenham, UK: Edward Elgar, pp. 201–215.

Håpnes, T. and K. H. Sørensen. 1995. Competition and collaboration in male shaping of computing: A study of a Norwegian hacker culture. In *The Gender-Technology Relation. Contemporary Theory and Research*, eds. K. Grint and R. Gill. London, UK: Taylor & Francis, pp. 174–191.

Håpnes, T. and B. Rasmussen. 1997. Internett—Jentenett? Ungdomsskolejenters databruk og datainteresser. Trondheim, Norway: Centre for women's studies, NTNU.

Håpnes, T. and B. Rasmussen. 2003. Gendering technology. Young girls negotiating ICT and gender. In *He, She and IT—Revisited*, ed. M. Lie. Oslo, Norway: Gyldendal Akademisk, pp. 173–197.

Hamilton, M. and B. Hokanson. 1999. Gender and software: Rhetoric and the computer culture exemplified by "Barbie Fashion Designer". In *Communication Jazz: Improvising the New International Communication Culture*, Proceedings from the 1999 IEEE International Professional Communication Conference, pp. 71–77.

Haraway, D. 1988. Situated knowledges: The science question in feminism and the privilege of partial perspective. *Feminist Studies*, 14 (3): 575–599.

Haraway, D. 1991. A cyborg manifesto: Science, technology and socialist-feminism in the late twentieth century. In *Simians, Cyborgs and Woman: The Reinvention of Nature*, ed. D. Haraway. London, UK: Routledge, pp. 149–181.

Harcourt, W, ed. 1999. *Women@internet. Creating new cultures in cyberspace.* London, UK: Zed books.

Harding, S. G. 1986. *The Science Question in Feminism.* Ithaca, NY: Cornell University Press.

Harris, K. and Dudley, M. 2000. Many First Steps: Interim Report to DTI and DfEE Learning Centre Board on the First Phase of "The Five" Pioneer Learning Centres, available at http://www.riservices.co.uk/Manyfirststeps.htm [accessed January 2004]

Hawthorne, S. and R. Klein, eds. 1999. *Cyberfeminism.* Melbourne, Victoria, Australia: Spiniflex.

Heimsvik, E., B. Nybakk and H. S. Spilker. 2005. Pandoraundersøkelsen 2005: Bruk av musikk og internett blant ungdom i Trondheim. Trondheim, Norway: Department of interdisciplinary studies of culture, NTNU.

Helsper, E. J. 2010. Gendered internet use across generations and life stages. *Communication Research*, 37 (3): 352–374.

Henning, A., A. Brenick, M. Killen, A. O'Connor and M. J. Collins. 2009. Do stereotypic images in video games affect attitudes and behavior? Adolescent perspectives. *Children, Youth and Environments*, 19 (1): 170–196.

Henwood, F. 1999. Exceptional women? Gender and technology in UK higher education. *IEEE Technology and Society Magazine*, 18 (4): 21–27.

Henwood, F. 2000. From the woman question in technology to the technology question in feminism: Rethinking gender equality in IT education. *The European Journal of Women's Studies*, 7 (2): 209–227.

Henwood, F., H. Kennedy and N. Miller. 2001. *Cyborg Lives: Women's Technobiographies*, York, UK: Raw Nerve.

Henwood, F., S. Plumeridge and L. Stepulevage. 2000. A tale of two cultures? Gender and inequality in computer education. In *Technology and In/equality: Questioning the Information Society*, eds. S. Wyatt, F. Henwood, N. Miller and P. Senker. London, UK: Routledge, pp. 111–128.

Herman, C. 2007. Public discourses private lives: East and West European Women's careers in Science, Engineering and Technology. Paper presented at 5th Gender Equality and Higher Education Conference, 28-31 August 2007, Humboldt University Berlin.

Hersh, M. 2000. The changing position of women in engineering world-wide. *IEEE Transactions of Engineering Management*, 47 (3): 345–359.

Himanen, P. 2001. *The Hacker Ethic and the Spirit of the Information Age.* New York: Random House.

Hofman, J. 1995. Writers, Texts and Writing Acts – Constructed Realities in Word Processing Software. Paper presented at the International Workshop –The Mutual Shaping of Gender and Technology, University of Twente.

Horowitz, R. ed, 2001. *Boys and Their Toys? Masculinity, Class and Technology in America.* London, UK: Routledge.

Houghton, D. J. and A. N. Joinson. 2010. Privacy, social network sites, and social relations. *Journal of Technology in Human Services*, 28: 74–94.

House, N. A. v. 2011. Feminist HCI meets Facebook: Performativity and social networking sites. *Interacting with Computers*, 23 (5): 422–429.

Jasanoff, S. 2004. Ordering knowledge, ordering society. In *States of Knowledge: The Co-Production of Science and Social Order*, ed. S. Jasanoff. London, UK: Routledge, pp. 13–45.

Jensen, K. S. H. 2005. Women Working in Science, Engineering and Technology: Higher Education and Industry. A literature review. The University of Salford. Available: www.cse.salford.ac.uk/dfe/literaturereviewMay.doc

Jenson, J. and S. d. Castell. 2008. Theorizing gender and digital gameplay: Oversights, accidents and surprises. *Eludamos. Journal for computer games culture*, 2 (1): 15–25.

Jenson, J. and S. d. Castell. 2011. Girls@Play. *Feminist Media Studies*, 11 (2): 167–179.

Jepson, A. and T. Perl. 2002. Priming the pipeline. *SIGSCE Bulletin*, 34 (2): 36–40.

Jorgenson, J. 2002. Engineering selves: Negotiating gender and identity in technical work. *Management Communication Quarterly*, 15 (3): 350–380.

Kafai, Y. B., C. Heeter, J. Denner and J. Y. Sun, eds. 2008. *Beyond Barbie and Mortal Kombat. New Perspectives on Gender and Gaming.* Cambridge, MA: The MIT Press.

Kanter, R. M. 1977. *Men and Women of the Corporation.* New York: Basic Books.

Katz, J. E. and P. Aspden. 1997. Motivations for and barriers to internet usage: Results of a national public opinion survey. *Internet Research,* 7 (3): 170–188.

Katz, J. E. and R. E. Rice, eds. 2002. *Social Consequences of Internet Use.* Cambridge, MA: MIT Press.

Kelan, E. K. 2008. Emotions in a rational profession: The gendering of skills in ICT work. *Gender, Work, and Organization,* 15 (1): 49–70.

Kerr, A., J. Kücklich and P. Brereton. 2006. New media—new pleasures? *International Journal of Cultural Studies,* 9(1): 63–82.

Kiesler, S., L. Sproull and J. S. Eccles. 1985. Pool halls, chips, and war games: Women in the culture of computing. *Psychology of Women Quarterly,* 9 (4): 451–462.

Kim, M. M. 2002. Cultivating intellectual development: Comparing women-only colleges and coeducational colleges for education effectiveness. *Research in Higher Education,* 43 (4): 447–481.

Kirton, G. and G. Healy. 2004. Shaping union and gender identities: A case study of women-only trade union courses. *British journal of industrial relations,* 42 (2): 303–323.

Kleif, T. and W. Faulkner. 2003. "I'm no athlete [but] I can make this thing dance!"—Men's pleasures in technology. *Science, Technology, & Human Values,* 28 (2): 296–325.

Koput, K. W. and B. A. Gutek. 2010. *Gender Stratification in the IT Industry: Sex, Status and Social Capital.* Cheltenham, UK: Edward Elgar.

Kramer, P. E. and S. Lehman. 1990. Mismeasuring women: A critique of research on computer ability and avoidance. *SIGNS: Journal of women in culture and society,* 16 (1): 158–171.

Kumar, K. 1995. *From Post-Industrial to Post-Modern Society: New Theories of the Contemporary World.* Oxford, UK: Blackwell.

Kvaløy, K. 1999. Fortellinger om moderne, flinke, lekne unge jenters forhold til datateknologi, SKF-rapport 3/2000. Trondheim, Norway: Centre for women's studies, NTNU.

Kvande, E. and B. Rasmussen. 1989. Men, women and data systems. *European Journal of Engineering Education,* 14 (4): 369–379.

Lagesen, V. A. 2003. Advertising computer science to women. (Or was it the other way around?). In *He, she and IT—Revisited,* ed. M. Lie. Oslo, Norway: Gyldendal Akademisk, pp. 69–102.

Lagesen, V. A. 2005. Extreme make-over? The making of gender and computer science. Ph.D. thesis, STS-report 71/2005. Trondheim, Norway: Centre for Technology and Society, NTNU.

Lagesen, V. A. 2007. The strength of numbers: Strategies to include women into computer science. *Social Studies of Science,* 37 (1): 67–92.

Lagesen, V. A. 2008. A cyberfeminist utopia? Perceptions of gender and computer science among Malaysian computer science students. *Science, Technology, & Human Values,* 33 (1): 5–27.

Lagesen, V. A. and K. H. Sørensen. 2009. Walking the line? The enactment of the social/technical binary in software engineering. *Engineering Studies,* 1 (2): 129–149.

Landström, C. 2007. Queering feminist technology studies. *Feminist Theory*, 8 (1): 7–26.

Latour, B. 1987. *Science in Action*. Maidenhead, Berkshire, UK: Open University Press.

Latour, B. 2005. *Reassembling the Social: An Introduction to Actor-Network-Theory*. Oxford, UK: Oxford University Press.

Laurie, N., C. Dwyer, S. L. Holloway and F. M. Smith. 1999. *Geographies of New Femininities*. Harlow, Essex, UK: Longman.

Lee, L., W. Faulkner and C. Alemany. 2007. Report on the Identification and Evaluation of Good Practice, Prometea: Empowering Women Engineers in Industrial and Academic Research, Deliverable No 18; ISSTI Working Paper. Edinburgh, UK: University of Edinburgh.

Lee, L., W. Faulkner and C. Alemany. 2010. Turning good policies into good practice: Why is it so difficult? *International Journal of Gender, Science and Technology*, 2 (1): 90–99.

Lente, H. v. 1999. Multimedia in the Netherlands: An overview of actors, activities and issues. In *Europe Appropriates Multimedia: A Study of the National Uptake of Multimedia in Eight European Countries and Japan*, eds. R. Williams & R. S. Slack. Trondheim, Norway: Centre for technology and society, NTNU, pp. 305–332.

Lerman, N. E., R. Oldenziel and A. P. Mohun, eds. 2003. *Gender & Technology. A Reader*. Baltimore, MD: The Johns Hopkins University Press.

Levy, S. 1984. *Hackers: The Heroes of the Computer Revolution*. New York: Bantam.

Lewis, K., J. Kaufman, M. Gonzalez, A. Wimmer and N. Christakis. 2008. Tastes, ties, and time: A new social network dataset using Facebook.com. *Social Networks*, 30: 330–342.

Lie, M. 1998. Computer Dialogues: Technology, Gender and Change, Report 2/1998. Trondheim, Norway: Centre for women's studies, NTNU.

Lie, M. and B. Rasmussen. 1983. Kan "kontordamene" automatiseres? (Can 'office girls' be automated?). Trondheim, Norway: Institute for social research in industry, SINTEF.

Lie, M., and K. H. Sørensen, eds. 1996a. *Making Technology Our Own? Domesticating Technology into Everyday Life*. Oslo, Norway: Scandinavian University Press.

Lie, M. and K. H. Sørensen. 1996b. Making technology our own? Domesticating technology into everyday life. In *Making Technology Our Own? Domesticating Technology into Everyday Life*, eds. M. Lie and K. H. Sørensen. Oslo, Norway: Scandinavian University Press, pp. 1–30.

Liff, S. 2002. Digital/Social: Inclusion/Exclusion, unpublished paper. Coventry, UK: Warwick Business School.

Liff, S. and I. Cameron. 1997. Changing equality cultures to move beyond "women's problems". *Gender, Work and Organization*, 4 (1): 35–46.

Light, J. S. 2001. Rethinking the digital divide. *Harvard Educational Review*, 71 (4): 709–733.

Light, A. 2011. HCI as Heterodoxy: Technologies of identity and the queering of interaction with computers. *Interacting with Computers*, 23: 430–438.

Littleton, K. and C. Hoyles. 2002. The gendering of information technology. In

Ghosts in the machine, eds. N. Yelland and R. Andee. New York: Peter Lang, pp. 3–32.

Livingstone, D. W. 2001. Adults' informal learning: definitions, findings, gaps and future research. NALL Working Paper #21-2001. Available: https://tspace.library.utoronto.ca/handle/1807/2735

Livingstone, S. and M. Bovill. 1999. Children, Young People, and the Changing Media Environment. London, UK: London School of Economic and Political Science.

Livingstone, S. and E. Hesper. 2007. Gradations in digital inclusion: Children, young people and the digital divide. *New Media & Society*, 9 (4): 671–696.

Lloyd, G. 1984. *The Man of Reason: "Male" and "Female" in Western Philosophy*. London, UK: Methuen.

Lovell, T. 1980. *Pictures of Reality: Aesthetics, Politics, Pleasure*. London, UK: British Film Institute.

Maccoby, E. E. and C. N. Jacklin. 1974. *The Psychology of Sex Difference*. Palo Alto, CA: Stanford University Press.

Mackay, H., C. Carne, P. Beynon-Davies and D. Tudhope. 2000. Reconfiguring the user: Using rapid application development. *Social Studies of Science*, 30 (5): 737–757.

Mahony, K. and B. Van Toen. 1990. Mathematical formalism as a means of occupational close computing—Why "hard" computing tends to exclude women. *Gender and Education*, 2 (3): 319–331.

Margolis, J. and A. Fisher. 2002. *Unlocking the Clubhouse: Women in Computing*. Cambridge, MA: MIT Press.

Margolis, J., A. Fischer and F. Miller. 2000. Caring about connections: Gender and computing. *IEEE Technology & Society Magazine*, 18 (4): 13–20.

Martinson, A., N. Schwartz and M. Walker Vaughan. 2002. Women's experience of leisure: Implication for design. *New Media and Society*, 4 (1): 29–49.

Mazman, S. G. and Y. K. Usluel. 2011. Gender differences in using social networks. *The Turkish Online Journal of Educational Technology*, 10 (2): 133–139.

McCarthy, H. 2004a. *Girlfriends in High Places*. London, UK: Demos.

McCarthy, H. 2004b. Old boys and new girls. In *Network Logic. Who governs in an interconnected world*, eds. H. McCarthy, P. Miller and P. Skidmore. London, UK: Demos, pp. 115–128.

McGivney, V. 1998. *Excluded Men: Men who are Missing from Education and Training*. Leicester, UK: National Institute of Adult Continuing Education.

McGivney, V. 1999. *Informal Learning in the Community: A Trigger for Change and Development*. Leicester, UK: National Institute of Adult Continuing Education.

McGivney, V. 2000. *Recovering Outreach: Concepts, Practices and Issues*. Leicester, UK: National Institute of Adult Continuing Education.

McIlwee, J. S. and J. G. Robinson. 1992. *Women in Engineering: Gender, Power, and Workplace Culture*. Albany, New York: State University of New York Press.

McKinney, V., D. D. Wilson, N. Brooks, A. O'Leary-Kelly and B. Hardgrave. 2008. Women and men in the IT profession. *Communications of the ACM*, 51 (2): 81–84.

McRobbie, A. 1991. *Feminism and Youth Culture: From Jackie to Just Seventeen.* London, UK: MacMillan.

Meiksins, P. and P. Whalley. 2002. *Putting Work in Its Place: A Quiet Revolution.* Ithaca, New York: Cornell University Press.

Mellström, U. 1995. Engineering Lives: Technology, Time and Space in a Male-Centred World. Linköping, Sweden: Linköping studies in arts and sciences.

Mellström, U. 2003. *Masculinity, Power and Technology: A Malaysian Ethnography.* Aldershot,Hampshire, UK: Ashgate.

Mellström, U. 2009. The intersection of gender, race and cultural boundaries, or Why is computer science in Malaysia dominated by women? *Social Studies of Science*, 39: 885–907.

Merchant, C. 1980. *The Death of Nature: Women, ecology and the scientific revolution.* San Francisco, CA: Harper & Row.

Mick, D. G. and S. Fournier. 1998. Paradox of technology: Consumer cognizance, emotions and coping strategies. *Journal of Consumer Research*, 25: 123–143.

Miles, I. and G. Thomas. 1996. User resistance to new interactive media: Participants, processes and paradigms. In *Resistance to New Technology*, ed. M. Bauer. Cambridge, UK: Cambridge University Press, pp. 255–275.

Millar, J. and N. Jagger. 2001. Women in ITEC Courses and Careers. Women and Equality Unit. Great Britain, UK: Department of Trade and Industry.

Miller, D. 2011. *Tales from Facebook.* Cambridge, UK: Polity.

Misa, T. J. ed. 2010. *Gender Codes: Why Women Are Leaving Computing.* Hoboken, NJ: Wiley.

Mörtberg, C. 1987. Varför har programmeryrket blivit mannligt? TULEA 1987:042. Luleå, Sweden: Luleå Technical University.

Mosco, V. 2004. *The Digital Sublime: Myth, Power and Cyberspace.* Cambridge, MA: MIT Press.

Murdock, G., P. Hartman and P. Gray. 1992. Contextualizing home computing. In *Consuming Technologies*, ed. R. Silverstone and E. Hirsch. London, UK: Routledge, pp. 146–160.

Newman, L. S., J. Cooper and D. N. Ruble. 1995. Gender and computers II: The interactive effects of knowledge and constancy on gender-stereotyped attitudes. *Sex roles: A Journal of Research*, 33 (5-6): 325–351.

Ng, C. and S. Mitter, eds. 2005. *Gender and the Digital Economy: Perspectives from the Developing World.* Thousand Oaks, CA: Sage.

Noble, D. F. 1992. *A World Without Women: The Christian Clerical Culture of Western Science.* New York: Alfred A. Knoph.

Nordli, H. 1998. Fra Spice Girls til Cyber Girls? En kvalitativ studie av datafascinerte jenter i ungdomsskolen, STS-report 35/1998. Trondheim, Norway: Centre for Technology and Society, NTNU.

Nordli, H. 2001. From Spice Girls to Cyber Girls: The role of multimedia in the construction of young girls' fascination for and interest in computers. In *Social Learning Technologies: The Introduction of Multimedia in Education*, ed. M. van Lieshout, T. Egyedi and W. E. Bijker. Aldershot, Hampshire, UK: Ashgate, pp. 110–133.

Nordli, H. 2003. The net is not enough. Searching for the female hacker, Ph.D. thesis, STS report 61/2003. Trondheim, Norway: Centre for Technology and Society, NTNU.

Norris, P. 2001. *Digital Divide. Civic Engagement, Information Poverty, and the Internet Worldwide.* West Nyack, New York: Cambridge University Press.

OECD. 2005. *Guide to Measuring the Information Society.* DSTI/ICCP/IIS (2005)6/FINAL.

OECD. 2007. *ICTs and Gender.* DSTI/ICCP/IE(2006)9/FINAL.

Offenbeek, M. A. G. v. and P. L. Koopman. 1996. Information system development: From user participation to contingent interaction among involved parties. *European Journal of Work and Organizational Psychology,* 5 (3): 421–438.

Oldenziel, R. 1999. *Making Technology Masculine: Men, Women and Modern Machines in America 1870-1945.* Amsterdam, the Netherlands: Amsterdam University Press.

Oldenziel, R. 2001. Man the maker, woman the consumer: The consumption-junction revisited. In *Feminism in Twentieth-Century Science, Technology, and Medicine,* eds. A. N. Creager, E. Lunbeck and L. Schiebinger. Chicago, IL: The University of Chicago Press, pp. 128–148.

Oliver, P. E. and G. Marwell. 2001. Whatever happened to critical mass theory? A retrospective and assessment. *Sociological Theory,* 19 (3): 293–311.

ONS. 2002. Labour Force Survey: Occupational Analysis, Women in IT. Spring Quarter. (UK) Office of National Statistics.

Otnes, C., T. M. Lowrey, and L. J. Shrum. 1997. Toward an understanding of consumer ambivalence. *Journal of Consumer Research,* 24: 81–93.

Oudshoorn, N., M. Brouns and E. v. Oost. 2006. Diversity and distributed agency in the design and use of medical video-communication technologies. In *Inside the Politics of Technology,* ed. H. Harbers. Amsterdam, the Netherlands: Amsterdam University Press, pp. 85–108.

Oudshoorn, N. and T. Pinch, eds. 2004. *How Users Matter: The Co-Construction of Users and Technologies.* Cambridge, MA: MIT Press.

Oudshoorn, N., E. Rommes and M. Stienstra. 2004. Configuring the user as everybody: Gender and design in information and communication technologies. *Science, Technology, & Human Values,* 29 (1): 30–63.

Oxaal, Z. and S. Baden. 1997. Gender and Empowerment: Definitions, Approaches and Implications for Policy. Brighton, UK: Institute of development studies, University of Sussex

Pain, D., J. Owen, I. Franklin and E. Green. 1993. Human-centred systems design: A review of trends within the broader systems development context. In *Gendered by Design,* eds. E. Green, A. Adam and D. Pain. London, UK: Taylor & Francis, pp. 11–30.

Perrons, D. 2003. The new economy and the work-life balance: Conceptual exploration case study of new media. *Gender, Work and Organisation,* 10 (1): 65–93.

Phipps, A. 2008. *Women in Science, Engineering and Technology: Three Decades of UK Initiatives.* Staffordshire, UK: Trentham Books.

Pitt, L. 2003. Masculinities@work: Gender inequalities in the new media industries. *Feminist Media Studies,* 3(3): 378–381.

Plant, S. 1997. *Zeroes and Ones: Digital Women and the New Technoculture.* London, UK: Doubleday and Forth Estate.

Plant, S. 2000. On the matrix: Cyberfeminist simulations. In *The Gendered*

Cyborg. A Reader, ed. G. Kirkup, L. K. Woodward and F. Hovenden. London, UK: Routledge, pp. 265–275.

Preston, P. 2005. ICTs in everyday life: Public policy implications for "Europe's way to the Information Society". In *Media Technology and Everyday Life in Europe: From Information to Communication*, ed. R. Silverstone. Aldershot, Hampshire, UK: Ashgate, pp. 195–212.

Preston, P. and C. MacKeogh. 2005. A "chilly culture" for women?: Gendered exclusions in ICT Work. In *The New World of Work: Labour Markets in Contemporary Ireland*, eds. G. Boucher and G. Colllins. Dublin, Ireland: The Liffey Press, pp.105–124.

Probert, B. and B. W. Wilson. 1993. *Pink Collar Blues: Work, Gender and Technology.* Melbourne, Victoria, Australia: Melbourne University Press.

Rana, S. 2008. Fulfilling technology's promise: Enforcing the rights of women caught in the global high-tech underclass. In *Women, Science and Technology: A Reader in Feminist Science Studies*, eds. M. Wyer, M. Barbercheck, D. G. Cockmeyer, H. Ö. Özturk and M. Wayne. London, UK: Routledge, pp. 322–342.

Rasmussen, B. 1997. Girls and computer science: It's not me. I'm not interested in sitting behind a machine all day. In *Women, Work and Computerisation. Spinning a Web from the Past to the Future*, ed. F. Grundy, D. Köhler, V. Oechtering and U. Petersen. Berlin, Germany: Springer, pp. 379–386.

Rasmussen, B. and T. Håpnes. 1991. Excluding women from the technologies of the future? A case study of the culture of computer science. *Futures*, 23: 1107–1119.

Rasmussen, B. and T. Håpnes. 2003. Gendering technology: Young girls negotiating ICTand gender. In *He, She and IT Revisited; New Perspectives on Gender in the Information Society*, ed. M. Lie. Oslo, Norway: Gyldendal Akademisk, pp. 173–197.

Ray, S. G. 2004. *Gender Inclusive Game Design: Expanding the Market.* Hingham, MA: Charles River Media.

Richardson, I. and I. Kavanagh. 1997. Positive action: Promoting technology and science through female role models. In *Women and Computing*, eds. R. Lander and A. Adam. Exeter, UK: Intellect.

Ridgeway, C. L. and S. J. Correl. 2004. Unpacking the gender system: A theoretical perspective on gender beliefs and social relations. *Gender & Society*, 18 (4): 510–531.

Roberts, E., M. Kassiandou and L. Irani. 2002. Encouraging women in computer science. *SIGCSE Bulletin*, 34 (2): 84–88.

Robertson, M., S. Newell, J. Swan, L. Mathiassen and G. Bjerknes. 2001. The issue of gender within computing: Reflections from the UK and Scandinavia. *Information Systems Journal*, 11: 111–126.

Robinson, G. J. and J. S. McIlwee. 1991. Men, women, and the culture of engineering. *The Sociological Quarterly*, 32 (3): 403–421.

Rogers, E. M. 1985. The diffusion of home computers among households in Silicon Valley. *Marriage and Family Review*, 8: 89–101.

Rogers, E. M. 2003. *Diffusion of Innovations*, 5th ed, New York: The Free Press.

Rommes, E. 2002a. Creating places for women on the Internet – The creation of a "Women's Square" in a digital city. *The European Journal of Women's Studies*, 9 (4): 401–427.

Rommes, E. 2002b. Worlds apart: Exclusion-processes in DDS. In *Digital Cities II; Computational and Sociological Approaches*, eds. M. Tanabe, P. v. d. Besselaar and T. Ishida, Berlin, Germany: Springer, pp. 219–232.

Rommes, E. 2002c. Gender Scripts and the Internet: The Design and Use of Amsterdam's Digital City. Enschede, the Netherlands: Twente University.

Rommes, E. W. M. 2010. Heteronormativity revisited: Adolescents' educational choices, sexuality and soaps. In *Gender Issues in Learning and Working with Information Technology: Social Constructs and Cultural Contexts*, eds. S. Booth, S. Goodman and G. Kirkup. Hershey, New.York: Information Science Reference, pp. 150–172.

Rommes, E., M. Bos and J. O. Geerdink. 2010. Design and use of gender specific and stereotypical toys. *International Journal of Gender, Science and Technology*, 3 (1): 186–204.

Rommes, E., W. Faulkner and I. van Slooten. 2005. Changing lives: The case for women-only vocational training revisited. *Journal of Vocational Education and Training*, 57 (3): 293–317.

Rommes, E., G. Overbeek, R. C. M. E. Engels, R. A. T. de Kemp and R. H. J. Scholte. 2007. "I'm not interested in computers". Gender-based occupational choices of teenagers. *Information, Communication & Society*, 10 (3): 299–319.

Rossiter, M. 1982. *Women Scientists in America: Struggles and Strategies to 1940.* Baltimore, MD: Johns Hopkins University Press.

Rossiter, M. 1995. *Women Scientists in America: Before Affirmative Actions 1940-1972.* Baltimore, MD: Johns Hopkins University Press.

Salminen-Karlsson, M. 1999. Bringing women into computer engineering. Curriculum reform processes at two institutes of technology. Ph.D. thesis, Linköping, Sweden: Linköping Studies in Education and Psychology.

Sassi, S. 2005. Cultural differentiation or social segregation? Four approaches to the digital divide. *New Media & Society*, 7 (5): 684–700.

Schiebinger, L, ed. 2008. *Gendered Innovations in Science and Engineering.* Palo Alto, CA: Stanford University Press.

Scott, A. 2001. (In)forming politics: Processes of feminist activism in the information age, *Women's Studies International Forum*, 24 (3/4): 409–421.

Scott, A. and M. Page. 2001. Change agency and women's learning: New practices in community informatics. In *Community Informatics. Shaping Computer-Mediated Social Relations*, eds. L. Keeble and B.D. Lauder. London, UK: Routledge.

Selwyn, N. 2004. Reconsidering political and popular understandings of the digital divide. *New Media & Society*, 6 (3): 341–362.

Shade, L. R. 2000. Courting women @ E-Com. In *Women, Work and Computerization, Charting a Course to the Future*, eds. E. Balka and R. Smith. Boston, MA: Kluwer Academic Publishers, pp. 217–224.

Siann, G. 1997. We can, we just don't want to. In *Women and Computing*, eds. R. Lander and A. Adam, Exeter, UK: Intellect.

Silverstone, R. 1994. *Television and Everyday Life.* London, UK: Routledge.

Silverstone, R., E. Hirsch and D. Morley. 1992. Information and communications technologies and the moral economy of the household. In *Consuming Technologies: Media and Information in Domestic Spaces*, eds. R. Silverstone and E. Hirsch. London, UK: Routledge, pp. 1–14.

Silverstone, R. and K. H. Sørensen. 2005. Towards "The Communication Society." In *Media, Technology and Everyday Life in Europe: From Information to Communication*, ed. R. Silverstone, Aldershot, Hampshire, UK: Ashgate, pp. 213–222.

Singh, K., K. R. Allen, R. Sheckler and L. Darlington. 2007. Women in computer-related majors: A critical synthesis of research and theory from 1994 to 2005. *Review of Educational Research*, 77 (4): 500–533.

Simpson, R. 2004. Masculinity at work: The experiences of men in women dominated occupations. *Work, Employment and Society*, 18 (2): 349–368.

Solomon, G., N. J. Allen and P. Resta, eds. 2003. *Toward digital equity: Bridging the divide in education.* Boston, MA: Allyn and Bacon.

Sørensen, K. H. 1992. Towards a feminized technology? Gendered values in the construction of technology. *Social Studies of Science*, 22: 5–31.

Sørensen, K. H. 2006. Domestication: The enactment of technology. In *Domestication of Media and Technology*, eds. T. Berker, M. Hartmann, Y. Punie and K.J. Ward, Maidenhead, Berkshire, UK: Open University Press, pp. 40–61.

Sørensen, K. H. and A.-J. Berg. 1987. Genderization of technology among Norwegian engineering students. *Acta sociologica*, 30 (2): 151–171.

Sørensen, K. H. and H. Nordli. 2005. Mobil moral og kjønn i endring? Mobiltelefon og hverdagsliv blant norske voksne. *Kvinneforskning*, 29 (1): 57–72.

Spender, D. 1995. *Nattering on the Net: Women, Power and Cyberspace.* North Melbourne, Victoria, Australia: Spinifex Press.

Spertus, E. 1991. Why Are There So Few Female Computer Scientists? MIT Artificial Intelligence Laboratory Technical Report 1315. Cambridge, MA: Massachusetts Institute of Technology.

Spilker, H. and K. H. Sørensen. 2000. A ROM of one's own or a home for sharing? Designing inclusion of women in multimedia. *New Media & Society*, 2 (3): 268–285.

Spilker, H. and K. H. Sørensen. 2002. Feminism for profit? Public and private gender politics in multimedia. In *Shaping Technology/Guiding Policy: Concepts, Spaces and Tools*, eds. K. H. Sørensen. and R. Williams. Cheltenham, UK: Edward Elgar, pp. 243–263.

Sproull, L., S. Kiesler and D. Zubrow. 1987. Encountering an alien culture. In *Computing and Change on Campus*, eds. S. B. Kiesler and L. S. Sproul. West Nyack, New York: Cambridge University Press, pp. 173–194.

Squires, J. 2005. Is mainstreaming transformative? Theorizing mainstreaming in the context of diversity and deliberation. *Social Politics*, 12 (3): 366–388.

Stanworth, C. 2000. Women and work in the information age. *Gender, Work and Organization*, 7 (1): 20–32.

Star, S. L. 1995. Introduction. In *The Cultures of Computing*, ed. S. L. Star. Oxford, UK: Blackwell Publishers, pp. 1–28.

Stepulevage, L. and S. Plumeridge. 1998. Women taking positions within computer science. *Gender and Education*, 10 (3): 313–326.

Stewart, J. 2002. Encounters with the information society: Personal and social issues in the appropriation of new media products in everyday life: Adoption, non-adoption, and the role of the informal economy and local experts. Ph.D. thesis. Edinburgh, UK: University of Edinburgh.

Stewart, J. 2003. The social consumption of information and communication technologies (ICTs): Insights from research on the appropriation and consumption of new ICTs in the domestic environment. *Cognition Technology and Work*, 5 (4): 4–14.

Stewart, J. 2007. Local experts in the domestication of information and communication technologies. *Information, Communication and Society*, 10(4): 547–569.

Stewart, J and S. Hyysalo. 2008. Intermediaries, users and social learning in technological innovation. *International Journal of Innovation Management*, 12(3): 295–325.

Stewart, J. and R. Williams. 2005. The wrong trousers? Beyond the design fallacy: Social learning and the user. In *Handbook of Critical Information Systems Research*, eds. D. Howcroft and E. M. Trauth. Cheltenham, UK: Edward Elgar, pp. 195–221.

Stratigaki, M. 2005. Gender mainstreaming vs. positive action. An ongoing conflict in EU gender equality policy. *European Journal of Women's Studies*, 12 (2): 165–186.

Studler, D. and I. McAllister. 2002. Does a critical mass exist? A comparative analysis of women's legislative representation since 1950. *European Journal of Political Research*, 41: 233–253.

Suchman, L. 1991. Closing remarks on the 4th conference on women, work and computerization: Identities and differences. In *Women, Work and Computerization*, eds. I. V. Eriksson, B. A. Kitchenham and K. G. Tijdens. Amesterdam, the Netherlands: Elsevier, pp. 431–437.

Suchman, L. 1994. Working relations of technology production and use. *Computer Supported Cooperative Work (CSCW)*, 2: 21–39.

Suchman, L. 2009. Agencies in technology design: Feminist reconfigurations. Paper presented at the Online Proceedings of the 5th European Symposium on Gender & ICT.

Sutton, R. E. 1991. Equity and computers in the schools—A decade of research. *Review of Educational Research*, 61 (4): 475–503.

Symmonds, J. 2000. Why I.T. doesn't appeal to young women. In *Women, Work and Computerization: Charting a Course to the Future*, eds. E. Balka and R. Smith. Boston, MA: Kluwer, pp. 70–77.

Taylor, T. L. 2003. Multiple pleasures: Women and online gaming. *Convergence: The journal of research into new media technologies*, 9 (1): 21–46.

Taylor, T. L. 2006. *Play Between Worlds. Exploring Online Game Culture*. Cambridge, MA: The MIT Press.

Teague, J. 2000. Women in computing: What brings them to it, what keeps them in it? *GATES*, 5 (1): 45–59.

Tett, L. 1994. Where have all the men gone? Adult participation in continuing education, *Scottish Journal of Adult and Continuing Education*, 1 (2): 41–49.

Thomas, G. and S. Wyatt. 2000. Access is not the only problem: Using and controlling the Internet. In *Technology and In/equality. Questioning the information society*, eds. S. Wyatt, F. Henwood, N. Miller and P. Senker. London, UK: Routledge, pp. 21–45.

Tonso, K. L. 1997. Constructing engineers through practice: Gendered features of learning and identity development. Ph.D. thesis. Boulder, CO: University of Colorado.

Tonso, K. L. 1999. Engineering gender—Gendering engineering: A cultural model for belonging. *Journal of Women and Minorities in Science and Engineering*, 5 (4): 365–404.

Tonso, K. L. 2006. Student engineers and engineer identity: Campus engineer identities as figured world. *Cultural Studies of Science Education*, 1 (2): 273–307.

Tonso, K. L. 2007. *On the Outskirts of Engineering: Learning Identity, Gender, and Power via Engineering Practice.* Rotterdam, the Netherlands: Sense Publishers.

Townsend, G. C. 2002. People who make a difference: Mentors and role models. *SIGSCE Bulletin*, 34 (2): 57–62.

Tsatsou, P. 2011. Digital divides revisited: What is new about divides and their research? *Media, Culture & Society*, 33 (2): 317–331.

Turkle, S. 1984. *The Second Self: Computers and the Human Spirit.* New York: Simon and Schuster.

Turkle, S. 1988. Computational reticence; why women fear the intimate machine. In *Technology and Women's Voices: Keeping in Touch*, ed. C. Kramarae. London, UK: Routledge and Kegan Paul, pp. 41–61.

Turkle, S. 2011. *Alone Together. Why We Expect More From Technology and Less from Each Other.* New York: Basic Books.

Vaage, O. 2002. *Norsk mediebarometer 2001*, Oslo, Norway: Statistics Norway.

Valenduc, G. et al. 2004. *Widening Women's Work in Information, Communication and Technology Work.* Namur, Belgium: Fondation Travail-Université.

van Emmerick, I. J. H. 2004. The more you can get the better: Mentoring constellations and intrinsic career success. *Career Development International*, 9 (6): 578–594.

van Zoonen, L. 2002. Gendering the internet. Claims, controversies and cultures. *European Journal of Communication*, 17 (1): 5–23.

Varma, R. 2002. Women in information technology: A case study of undergraduate students in a minority-serving institution. *Bulletin of Science, Technology and Society*, 22: 274–282.

Volman, M. and E. van Eck. 2001. Gender equity and information technology in education: The second decade. *Review of Educational Research*, 71 (4): 613–634.

Volman, M., E. van Eck, I. Heemskerk and E. Kuiper. 2005. New technologies, new differences. Gender and ethnic differences in pupils' use of ICT in primary and secondary education. *Computers & Education*, 45 (1): 35–35.

Wajcman, J. 1991. *Feminism Confronts Technology.* Cambridge, UK: Polity Press.

Wajcman, J. 2004. *Technofeminism.* Cambridge, UK: Polity Press.

Wajcman, J. 2009. Feminist theories of technology. *Cambridge Journal of Economics*, 34 (1): 143–152.

Walby, S. 2005. Gender mainstreaming: Productive tensions in theory and practice. *Social Politics*, 12 (3): 321–343.

Walsh, V., R. Roy, M. Bruce and S. Potter. 1992. *Winning by Design: Technology, Product Design and International Competitiveness.* Oxford, UK: Blackwell Business.

Walther, J. B., B. v. d. Heide, S.-Y. Kim, D. Westerman and S. T. Tong. 2008. The role of friends' appearance and behaviour on evaluations of individuals on

Facebook: Are we known by the company we keep? *Human Communication Research*, 34, 28–49.

Webster, F, ed. 2004. *The Information Society Reader*. London, UK: Routledge.

Webster, J. 1996. *Shaping Women's Work. Gender, Employment and Information Technology*. London, UK: Longman.

Webster, J. 2005. Why are women still so few in IT? Understanding the persistent under-representation. In *The Gender Politics of ICT*, ed. J. Archibald, J. Emms, F. Grundy, J. Payne and E. Turner. London, Middlesex, UK: Middlesex University Press, pp. 3–14.

Weinberger, C. J. 2004. Just ask! Why surveyed women did not pursue IT courses or careers. *IEEE Technology and Society Magazine*, 23 (2): 28–35.

Weingart, P. 1989. Grosstechnische Systeme—Ein Paradigma der Verknüpfung von Technikentwicklung und sozialem Wandel. In *Technik als sozialer prozess*, ed. P. Weingart. Berlin, Germany: Suhrkamp, pp. 174–196.

Weizenbaum, J. 1976. *Computer Power and Human Reason: From Judgment to Calculation*. San Francisco, CA: W. H. Freeman.

West, C. and D. H. Zimmerman. 1987. Doing gender. *Gender & Society*, 1 (2): 125–151.

West, M. and Ross, S. 2002. Retaining females in computer science: A new look at a persistent problem. *Journal of computing sciences in colleges*, 17 (5): 1–7.

Whyte, J. 1986. *Girls into Science and Technology: The Story of a Project*. London. UK: Routledge and Kegan Paul.

Wigfield, A. and Eccles, J.S. 2000. Expectancy value theory of achievement motivation. *Contemporary Educational Psychology*, 25, 68–81.

Wilding, F. 1998. Notes on the political conditions of cyberfeminism. *Art Journal*, 57 (2): 46–59.

Williams, P. 1991. *The Alchemy of Race and Rights*. Cambridge, MA: Harvard University Press.

Williams, R., W. Faulkner and J. Fleck, eds. 1998. *Exploring Expertise: Issues and Perspectives*. London, UK: Macmillan.

Williams, R., Stewart, J and Slack, R. 2005. *Social Learning in Technological Innovation*. Cheltenham, UK: Edward Elgar.

Willis, P. 1977. *Learning to Labour*. Aldershot, UK: Gower.

Wilson, F. 2003. Can compute, won't compute: women's participation in the culture of computing'. *New Technology, Work and Employment*, 18 (2): 127–142.

Womeng. 2005. Creating Cultures of Success for Women Engineers. Womeng Synthesis Report, EC 5th FP, HPSE-CT-2002-00109.

Woodfield, R. 2000. *Women, Work and Computing*. Cambridge, UK: Cambridge University Press.

Wright, R. 1996. The occupational masculinity of computing. In *Masculinities in Organisations*, ed. C. Cheng, Thousand Oaks, CA: Sage, pp. 77–96.

Wyatt, S. 2004. Non-users also matter: The construction of users and non-users of the Internet. In *How Users Matter: The Co-construction of Users and Technology*, ed. N. Oudshoorn and T. J. Pinch. Cambridge, MA: MIT Press, pp. 67–80.

Wyatt, S. 2008. Feminism, technology and the information society: Learning from the past, imagining the future. *Information, Communication & Society*, 11 (1): 11–130.

Wyatt, S., F. Henwood, A. Hart and J. Smith. 2005. The digital divide, health information and everyday life. *New Media and Society*, 7: 199–218.

Wyatt, S., F. Henwood, N. Miller and P. Senker, eds. 2000. *Technology and In/equality. Questioning the information society*. London, UK: Routledge.

Xiao, L. and J.M. Carroll. 2007. Fostering an informal learning community of computer technologies at school. *Behaviour and Information Technology*, 26 (1): 23–36.

Yelland, N. and A. Rubin, eds. 2002. *Ghosts in the Machine: Women's Voices in Research with Technology*. New York: Peter Lang.

Authors

Dr. Wendy Faulkner

Wendy Faulkner was involved in early feminist studies of science and technology, in Alice Through the Microscope (Brighton Women and Science Group, 1980) and Smothered by Invention (with Erik Arnold, 1985). Her subsequent academic career took her into innovation studies and the sociology of technology, including ethnographic research on gender dynamics within engineering workplaces. She was involved in setting up and developing the University of Edinburgh Masters and Doctoral Programmes in Science and Technology Studies at the University of Edinburgh for over 20 years before retiring in 2009. She now offers training and consultancy on dialogic approaches to public engagement.

Dr. Helen Jøsok Gansmo

Helen Jøsok Gansmo is Senior research scientist and Associate Professor at the Department of interdisciplinary studies of culture at the Norwegian University of Science and Technology. Her field of research is concerned with social studies of technology and the making of everyday life. Past and current research projects are related to translations of gender ICT policies, planning of carbon neutral settlements, and changing energy use in everyday lives.

Dr. Lisa Lee

Lisa Lee is an Honorary Fellow at the Institute for the Study of Science, Technology and Innovation (ISSTI), University of Edinburgh. Her field of research is concerned with social studies of technology, and includes work on young people and the Internet, mobile and wireless technology use, gender inclusion in SET (Science, Engineering and Technology) education and employment, and most recently on resistance to technology through the production and consumption of home grown food.

Dr. Els Rommes

Els Rommes is assistant Professor at the Institute for Gender Studies, Radboud University Nijmegen. Her research and publication areas include gendered design and use of ICTs, representations of SET (science, engineering and technology) and teenagers' professional choices. She has participated in various projects, amongst others in the EU project 'Motivation; Promoting positive images of Science, Engineering and Technology in young People', and 'Necessity or Nerds; on representations of SET in the Dutch media'. She is (co)author of the books 'Gender Scripts and the Internet; The Design and Use of Amsterdam's Digital City' (2002); 'Gender Perspectives on Information Society Technology' (2007) and of numerous articles.

Professor Vivian Anette Lagesen

Vivian Anette Lagesen is Professor in Science and Technology Studies at the NTNU. Her research interests include gender and computer science and inclusion strategies in a comparative perspective, where she has published widely. She has also worked with knowledge management, learning, occupational culture and gender in the software industry and studied this in different countries. She has continued her interest with engineering studies, and has also been working with engineers and environmental knowledge as well as transition strategies and energy studies.

Professor Knut H. Sørensen

Knut H. Sørensen is Professor of Science and Technology Studies at Department of Interdisciplinary Studies of Culture, Norwegian University of Science and Technology in Trondheim. He has been involved in studies of gender, science and technology since the later 1970s. His other fields of research include studies of information and communication technology, domestication of technology, sustainable energy studies, climate issues, and interdisciplinarity. Sørensen has a wide experience teaching STS and has been responsible for developing the PhD programme of the department.